MUSCLE BIOLOGY

MUSCLE BIOLOGY

The Life History of a Muscle

BRUCE M. CARLSON

Professor Emeritus of Anatomy
Department of Cell and Developmental Biology
University of Michigan Medical School
Ann Arbor, MI
United States

Academic Press is an imprint of Elsevier
125 London Wall, London EC2Y 5AS, United Kingdom
525 B Street, Suite 1650, San Diego, CA 92101, United States
50 Hampshire Street, 5th Floor, Cambridge, MA 02139, United States
The Boulevard, Langford Lane, Kidlington, Oxford OX5 1GB, United Kingdom

Notices

Knowledge and best practice in this field are constantly changing. As new research and experience broaden our understanding, changes in research methods, professional practices, or medical treatment may become necessary.

Practitioners and researchers must always rely on their own experience and knowledge in evaluating and using any information, methods, compounds, or experiments described herein. In using such information or methods they should be mindful of their own safety and the safety of others, including parties for whom they have a professional responsibility.

To the fullest extent of the law, neither the Publisher nor the authors, contributors, or editors, assume any liability for any injury and/or damage to persons or property as a matter of products liability, negligence or otherwise, or from any use or operation of any methods, products, instructions, or ideas contained in the material herein.

Library of Congress Cataloging-in-Publication Data
A catalog record for this book is available from the Library of Congress

British Library Cataloguing-in-Publication Data
A catalogue record for this book is available from the British Library

ISBN: 978-0-12-820278-4

For information on all Academic Press publications visit our website at
https://www.elsevier.com/books-and-journals

Publisher: Andre Gerhard Wolff
Acquisitions Editor: Andre Gerhard Wolff
Editorial Project Manager: Barbara Makinster
Production Project Manager: Niranjan Bhaskaran
Cover Designer: Matthew Limbert

Typeset by TNQ Technologies

Contents

Sources of borrowed figures

Biral D, et al. Neurol Res 2008;30:140.
Blau H, et al. Nature Med 2015;21:856.
Boron WF, Boulpaep EL. Medical Physiology 2/e. Philadelphia: Elsevier Saunders; 2012.
Botinelli R, Reggiani C. Skeletal Muscle Plasticity in Health and Disease. Berlin: Springer; 2006.
Carlson BM. Principles of Regenerative Biology. San Diego: Academic Press; 2007.
Carlson BM. The Human Body – Linking Structure and Function. San Diego: Academic Press; 2019a.
Carlson BM. Human Embryology and Developmental Biology 6/e. Philadelphia: Elsevier; 2019b.
Carosio S, et al. Aging Res Revs 2011;10:35
Carroll RG. Elsevier's Integrated Physiology. Philadelphia: Mosby Elsevier; 2007.
Ciciliot S, Schiaffino S. Curr Pharmaceut Design 2010;16:909.
Clark B. Med Sci Sports Exerc 2009;41:1871.
Dubowitz V, et al. Muscle Biopsy: a Practical Approach 4/e. Philadelphia: Saunders Elsevier; 2013.
Dumont NA, et al. Compr Physiol 2015;5:1048.
Gartner LP, Hiatt JL. Color Textbook of Histology 3/e. Philadelphia: Saunders Elsevier; 2007
Goebel HH, et al. Muscle Diseases – Pathology and Genetics. London: Wiley Blackwell; 2013
Jorde LB, Carey JC, Bamshad MJ. Medical Genetics 4/e. Philadelphia: Mosby Elsevier; 2010
Kierszenbaum AL, Tres LL. Histology and Cell Biology 3/e. Philadelphia: Elsevier Saunders; 2012.
Kumar V. Robbins and Cortran Pathologic Basis of Disease 9/e. Philadelphia: Elsevier; 2015.
Larsson L, et al. Physiol Rev 2019;99:441.
LeBlanc A, et al. Int J Sports Med 1997;18:S283.
Mantilla CB, et al. Neuroscience 2007;146:181.
Myers TW. Anatomy Trains 3/e. Edinburgh: Churchill Livingstone Elsevier; 2014.
Navas-Enamorado I, et al. Aging Res Revs 2017;37:39.
Nolte J. The Human Brain 6/e. Philadelphia: Mosby Elsevier; 2009.
Piasecki M, et al. Biogerontology 2016;17:93.

Pollard TD, et al. Cell Biology. Philadelphia: Saunders; 2017.
Powers SK, Howley ET. Exercise Physiology 10/e. New York: McGraw-Hill; 2018.
Schiaffino S, et al. Physiol Rev 1996;76:371.
Schoenwolf GC, et al. Larsen's Human Embryology 5/e. Philadelphia: Elsevier Churchill Livingstone; 2015
Schuelke M, et al. New England J Med. 2004; 350:2682.
Stevens A, Lowe J. Human Histology 3/e. Philadelphia: Elsevier Mosby; 2005.
Thibodeau GA, Patton KT. Anatomy & Physiology 6/e. St. Louis: Mosby Elsevier; 2007.
Veeck LL, Zaninovic N. An Atlas of Human Blastocysts. Boca Raton: Parthenon; 2003.
Wosczyna MN, Rando TA. Devel Cell 2018;46:135.
Young B, Loew JS, Stevens A, Heath JW. Wheater's Functional Histology 5/e. Edinburgh: Churchill Livingstone Elsevier; 2006.

Preface

From its subtitle, one might expect that this book concerns the life history of a single muscle. However, shortly after starting research on the book, I realized that the published literature on any single muscle would by no means do justice to the topic. Therefore, I soon decided to broaden the scope to include any of the skeletal muscles in the body that are relevant to the topic at hand. Nevertheless, the purpose of the book remains the same, namely, to take the reader on a journey that begins with the first traces of muscle formation in the embryo and follows the growth, development, and adaptations of muscles through the entire life cycle until the end of life.

This book can be used as an introductory text to a course on muscle biology, but I hope that it can also be read in a more casual setting—maybe not as fast-moving as thriller novel, but yet resembling a story. Although subjective in the choice of material presented, it has been my intent to cover most of the major developmental events that could occur to a muscle over the course of its lifetime.

At a purely personal level, the time course of this book represents a summary of the many phases of my own career in muscle biology. I have been writing embryology textbooks since 1973 and have done some research on the embryology of limb muscles. Since the mid-1960s, muscle regeneration and the contractile physiology of regenerating and transplanted muscle have been a main focus of my research program. The last 2 decades of my research career concentrated on the biology of aging and long-term denervated muscle. Throughout the book, my bias as a developmental biologist, rather than a physiologist or biochemist, will be apparent. I have always viewed development as a continuum that begins with the fertilized egg and ends with death. This viewpoint represents the way in which the material in this book is packaged.

As is usual with my books, I welcome comments, especially if they point out the inevitable mistakes that can appear in a book. Feel free to contact me at brcarl@umich.edu.

Bruce Carlson

Introduction

Skeletal muscle is one of the most prominent tissues in the body. Almost invariably, when one thinks of muscles, some of the first images that come to mind are those of well-conditioned young athletes whose muscles are at the peak of their structural and functional capacities. Yet, this represents only a tiny slice of what it is like to be a muscle. Like every other part of the body, what one sees at any given moment represents a segment of a temporal continuum that begins with a fertilized egg and ends with death. From moment to moment, the body is never the same, and at all levels from the molecular to gross anatomical features, the body is constantly changing.

At the molecular level, countless biochemical reactions are constantly taking place within a time frame measured in milliseconds, whereas at the highest levels of organization, such as the heart or brain, some changes might not be apparent for months or even years, even though this seemingly glacial rate of change masks many events and modifications that are not apparent to the casual observer. During DNA replication, 50–100 base pairs are added every second at every replication fork, of which there almost 100,000 per chromosome. In the adult human, approximately 2.4 million new red blood cells are formed each second in order to replace those that die. At the other end of the spectrum, cardiomyocytes remain stable as cellular entities throughout one's lifespan, although internally the components turn over at an amazing rate. Between these extremes, gross hypertrophy of skeletal muscles during strength training begins to become apparent over the course of several weeks, and the turnover of Haversian systems in long bones takes place in roughly the same time frame.

For any organ in the body, its life history can be broken down into several general stages. Starting with the early embryo, at some point, a group of cells receives the information that unlocks their potential to form a specific type of tissue. This process is called specification or determination. Then, through mechanisms that remain remarkably poorly understood, these cells receive additional information that determines the overall form of the structure that they will become. This process is known as morphogenesis. As the organ is being put together within the embryo, the cells within it go through a process of differentiation, during which they become structurally and functionally specialized for fulfilling their mature function. In the human, these processes take place during the first 8 weeks of embryonic development. Cellular differentiation, however, continues over a longer period—often until after birth.

During the fetal period (from 3 to 9 months of pregnancy), morphogenesis is largely completed. Essentially, all organs and structures are recognizable as anatomical entities although their proportions are often far from those seen in the adult. Much of the fetal period is occupied by growth and functional development (both mechanical and biochemical) of tissues and organs. Birth itself has profound implications for the lungs and overall pattern of circulation, but for many other structures, the postnatal period represents a continuation of processes that were already underway in the fetus.

The period from birth until approximately the age of 20 is marked by rapid but decreasing rates of growth and functional development. Much of what is happening within the body follows an internal program, but at all times environmental influences that are superimposed upon these internal events can modify the course or amount of development.

Adulthood is characterized by the final maturation and maintenance of most structures and organs. Although developmentally this represents a period of considerable stability, it is also the period when environmental influences often exert a greater effect than internally directed changes.

Aging represents the last major phase of one's life cycle, but the onset of aging does not appear at a single time point. A good example is the process of accommodation in vision. Most people don't need bifocals in order to read until they are in their 40s, but yet, the aging changes in the lens apparatus of the eye begin at around age 12. It simply takes over 30 years before the body has lost the ability to adapt to these changes. Aging is a gradual process that gathers steam in one's late 50s and 60s. Over the succeeding decades, the process of aging accelerates, but often biological aging processes are masked by pathological changes.

Skeletal muscle follows all the above life cycle stages, and this text is organized around these stages. Commitment to becoming muscle cells mostly takes place during the fourth or fifth week of embryological development, and the earliest stages of muscle morphogenesis take place shortly before or after that time depending upon the muscle group. By the fetal stage at 9weeks, all muscles are identifiable, and rapid growth and the early stages of functional development are taking place.

The early postnatal period and childhood are dominated by the growth of the muscular system and its functional maturation. The latter is heavily influenced by maturational changes in the nervous system, especially myelinization of axons of motor tracts within the brain and spinal cord. Myelinization of some tracts is not completed until one's early 20s. Other internal influences on muscle development include endocrine effects. With the onset of puberty, the effects of the sex hormones—estrogens and testosterone—are highly evident on many parts of the body, especially the skin and hairs, but these hormones also exert a significant effect on skeletal muscle. In addition to these internal influences, the skeletal musculature of children and adolescents is highly responsive to training and other mechanical environmental influences. Many of the genetically -based muscle diseases, especially the muscular dystrophies, become evident early in childhood.

In adulthood, the muscular system is basically stable, but it continues to be highly responsive to the effects of increased use or underuse through mechanisms of hypertrophy or atrophy. Among physically active people, gross or micro muscle trauma is followed by a characteristic sequence of muscle fiber degeneration and regeneration. Depending upon the type of activity, muscle capacity peaks during one's 20s and 30s. Peak capacity then remains stable for varying periods of time and then begins a slow pathway of functional decline.

There is no exact timetable for the onset of muscular old age. Rather, it is a gradual process that can be significantly affected by the amount of use of the muscular system. In addition, the aging of both muscle structure and functional capacity becomes heavily dependent upon changes in the nervous system. The state of motor innervation, in particular, exerts a profound effect upon the muscle fibers that are innervated by individual motor nerve fibers. Aging changes affect not only the muscle fibers themselves, but also their connective tissue investments. Many of the muscle injuries that increase in number and intensity as aging progresses are actually injuries to the tendons and the connective tissue sheaths within the muscles. Studies of muscles in extremely old rats have shown what appears to be a loss of some of the overall homeostatic mechanisms that keep muscle tissue intact. Some phenomena that normally occur in the embryo reappear in very old muscle.

The overall approach to this book is to focus on skeletal muscle as a continuum, from its first appearance in the embryo, through later development to maturity, and then the slow decline as aging progresses. Each of these stages has its own unique characteristics. At any given stage of development, those influences that affect the structure and function of a muscle will be stressed. The goal of this book is to demonstrate that skeletal muscle is a highly dynamic tissue that responds not only to internal cues, but is also highly responsive to environmental conditions and functional demands. I hope that you, upon finishing this text, will come away with a much greater appreciation of the dynamic nature of muscle tissue.

CHAPTER 1

Mature skeletal muscle—An overview

The purpose of this chapter is to introduce some fundamental background on the structure and function of skeletal muscle that will be needed to fully appreciate the material covered in the succeeding chapters. For those not familiar with both the terminology and basic concepts of muscle organization and function, some of the following chapters, such as those on the embryonic development and early growth of muscle, would be a tough slog. Thus, we will break temporal continuity by beginning with an overview of how a mature skeletal muscle is put together and how it functions.

The many structural dimensions of muscle

A working understanding of skeletal muscle requires a knowledge of how muscles are put together and how they function at many levels—from the interactions of molecules to groups of muscles working in concert to produce specific movements of the body. Most of the 600+ muscles in the human body are built according to a couple of basic organizational plans. The most common, and characteristic of limb muscles, is a roughly cylindrical muscle with a well-defined tendon on either end that connects the muscle to a bone. The other plan, seen mainly in muscles of the trunk and head, consists of a more flattened muscle with at least one end connecting broadly to either bone or connective tissue through an **aponeurosis**. In contrast to a **tendon**, which concentrates the contractile force of a muscle onto a small point, an aponeurosis distributes the force of a muscle over a wide area.

The contractions of individual muscle fibers produce the forces necessary for the production of gross movements through the layers of connective tissue that invest them. These layers range from the **epimysium**, which envelops an entire muscle, to the **perimysium**, which surrounds bundles of muscle fibers (**fascicles**), to the wispy **endomysium**, associated with individual muscle fibers (Fig. 1.1).

Other gross features of muscles are nerves and blood vessels. In small muscles, they enter the muscle at a single main site, but larger muscles often have several neurovascular entry points. Such entry points include motor and sensory nerve fibers, small arteries and veins, and lymphatic channels. Within the muscle, successively smaller arterial branches spread out throughout the muscle until each muscle fiber is surrounded by a network of capillaries, which then drain into a parallel venous system that removes the blood from the muscle. Situated alongside the blood vessels, lymphatic channels also drain fluid from muscles. Lacking a muscular wall, the lymphatic channels rely upon muscle contractions to propel the lymph out from a muscle and into the larger lymphatic circulation.

Muscle Biology
ISBN 978-0-12-820278-4, https://doi.org/10.1016/B978-0-12-820278-4.00004-9

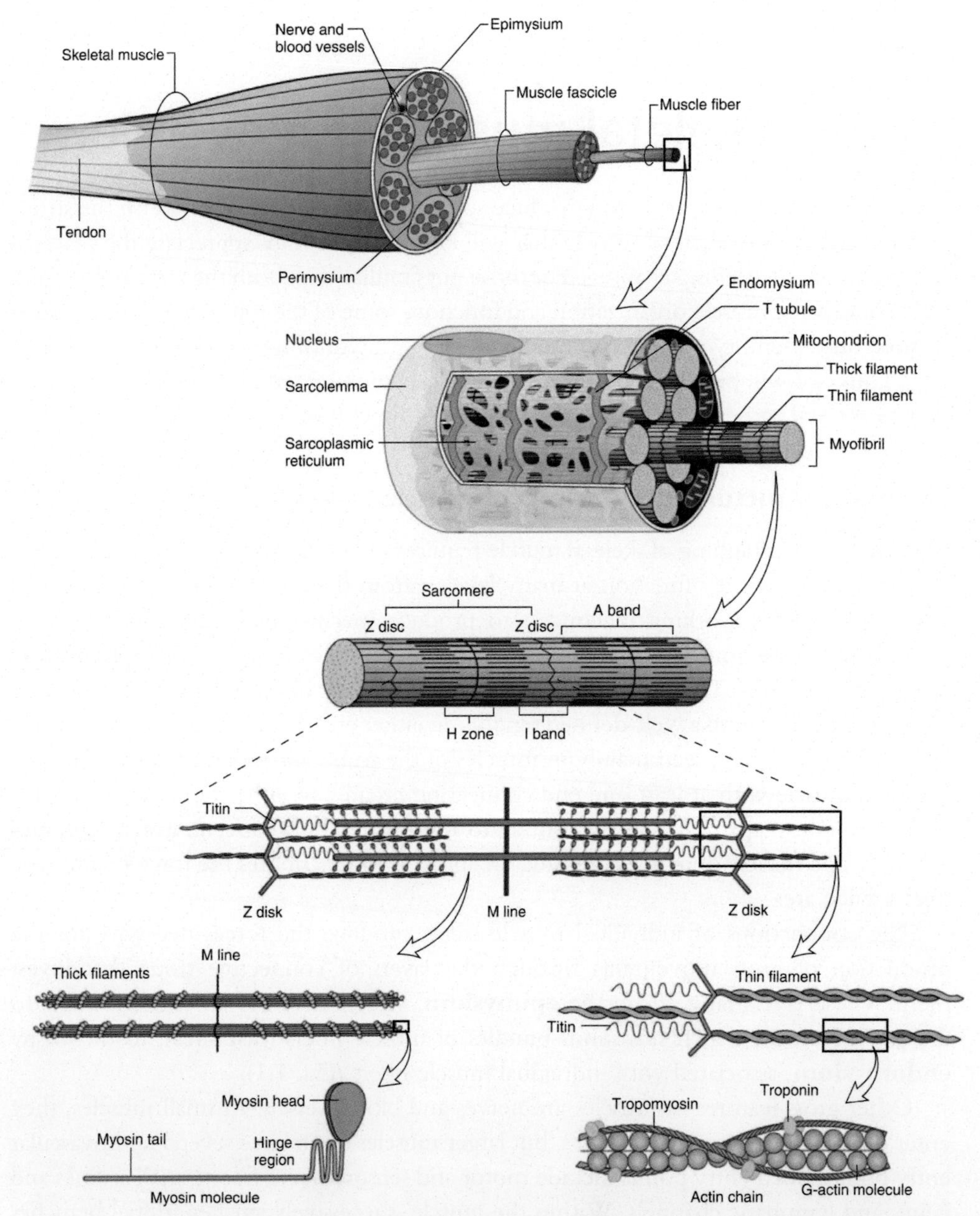

Figure 1.1 A scheme showing the various dimensions of a skeletal muscle from its gross structure to contractile molecules. *(From Carroll (2007), with permission.)*

Muscle fibers represent the fundamental contractile unit of a muscle. A muscle fiber is a long multinucleated cell, ranging from millimeters to many centimeters in length (up to 30 cm in the human sartorius muscle) and from tens to hundreds of μm in diameter. Upon contraction, a typical muscle fiber may shorten by up to 20% of its resting length. A mature muscle fiber contains hundreds of peripherally located nuclei that surround many bundles of highly organized contractile proteins. Each muscle fiber is closely invested by a **basal lamina**, which is produced by the muscle fiber itself. Between the basal lamina and the outer membrane of the muscle fiber (the **sarcolemma**[1]) is a small population of **satellite cells**, the stem cells that are responsible for both the growth and regeneration of individual muscle fibers (Fig. 1.2).

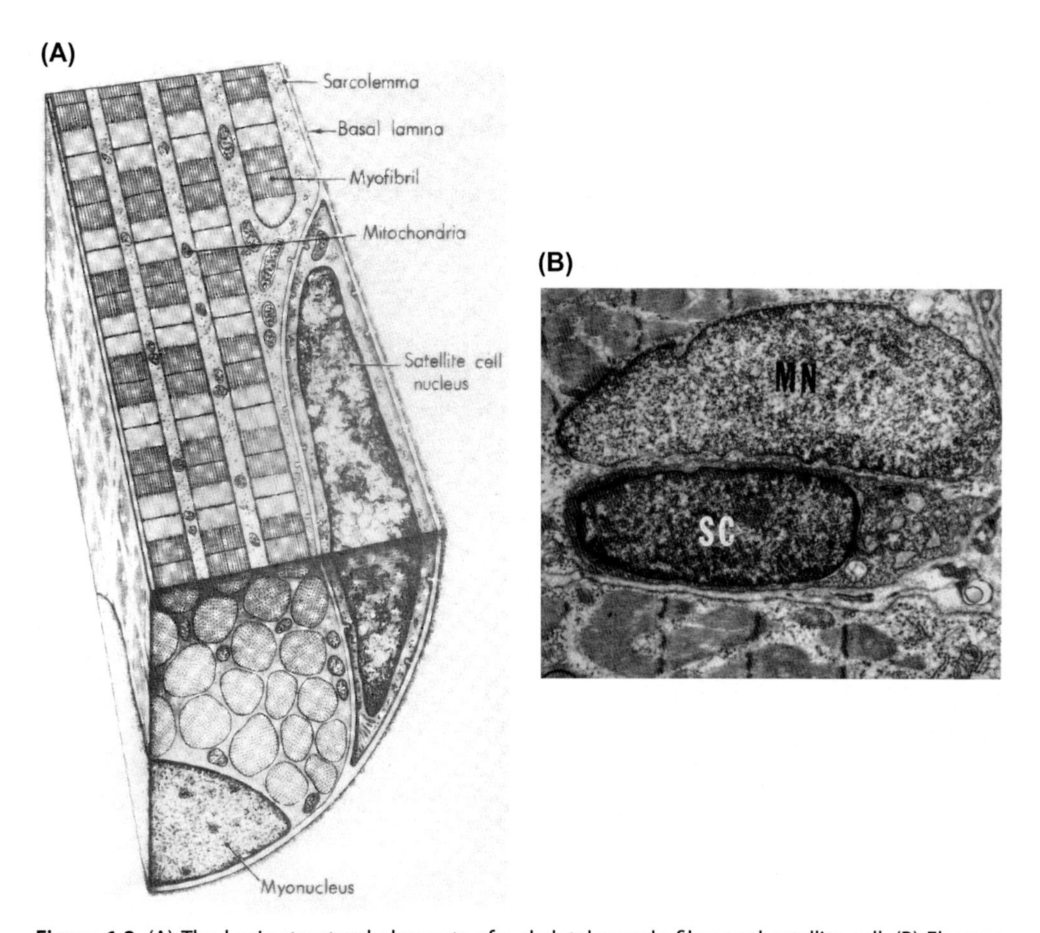

Figure 1.2 (A) The basic structural elements of a skeletal muscle fiber and satellite cell. (B) Electron micrograph of cross-sectioned muscle fiber and satellite cell. *MN*, Myonucleus; *SC*, satellite cell. *(Courtesy of M.H. Snow.)*

[1] The prefix sarc-appears in many words associated with skeletal muscle. It is derived from the Greek word, *sarx*, meaning flesh.

The interior of a muscle fiber consists of highly ordered arrays of contractile proteins and the associated molecules that keep them in close register. Even at the light microscopic level, this high degree of order gives muscle fibers a cross-striated appearance, hence the name **striated muscle** given to both skeletal and cardiac muscle (Fig. 1.3). The various components of the cross-striations were given names (and letter abbreviations) based on their optical properties, sometimes under polarized light. As seen in high-power electron micrographs, these components stand out clearly.

The fundamental contractile unit of a muscle fiber is a 2.3 μm segment called a sarcomere. **Sarcomeres** are arranged in serial fashion throughout the length of a muscle fiber. Each sarcomere is bounded on either end by narrow **Z lines** (see Fig. 1.3A and C). The banding patterns within a sarcomere are based upon the ordering of thick and thin filaments. **Thick filaments** are aggregates of individual **myosin** molecules. Each myosin molecule is pipe-shaped, with a long tail and a protruding head. Groups of myosin molecules making up a thick filament are arranged with their tails back-to-back, making the thick filaments bilaterally symmetrical (Fig. 1.4B). Within a sarcomere, the space from one end of a thick filament to the other is called the **A band** (A for anisotropic optical qualities). **Thin filaments** consist of a double helix of polymeric **actin** strands interwoven with filamentous **troponin** molecules that are periodically studded with **troponin** units (Fig. 1.5). Thick and thin filaments overlap to a certain extent, depending upon the degree of contraction of the muscle. This can be readily seen in cross section (see Fig. 1.4B). A lighter region (the **H** band [from the German *helle*, meaning light]) within the overall A band is the area occupied only by thick filaments. The parts of the sarcomere containing thin, but not thick filaments are called the **I bands** (I for isotropic).

Thick and thin filaments are held in register by a number of other proteins. The Z line is a meshwork of at least a half dozen proteins that anchor the ends of thin filaments from adjacent sarcomeres. At the very center of the sarcomere, another group of proteins, visible as the **M line**, stabilizes and spaces the thick filaments during various phases of muscle contraction. The regular lateral spacing of thin filaments is accomplished largely through a very large (up to 900 kD) filamentous protein called **nebulin**. **Titin**, an enormous filamentous protein, the largest protein in the body (3700 kD), extends from the Z band to the M band. In addition to helping to separate thick filaments, titin is highly elastic and provides passive resistance to stretching of the muscle fiber.

Within a muscle fiber, groups of contractile proteins are aggregated into parallel longitudinal bundles, called **myofibrils** (see Fig. 1.1). The sarcomeres of adjacent myofibrils are all kept in register by an intermediate filament protein called **desmin**, which connects z bands in a side-by-side fashion. Overall, the tight organization, both longitudinally and laterally, between the thick and thin filaments and their connecting proteins allows the molecular interactions between actin and myosin that produce the force in a contracting muscle.

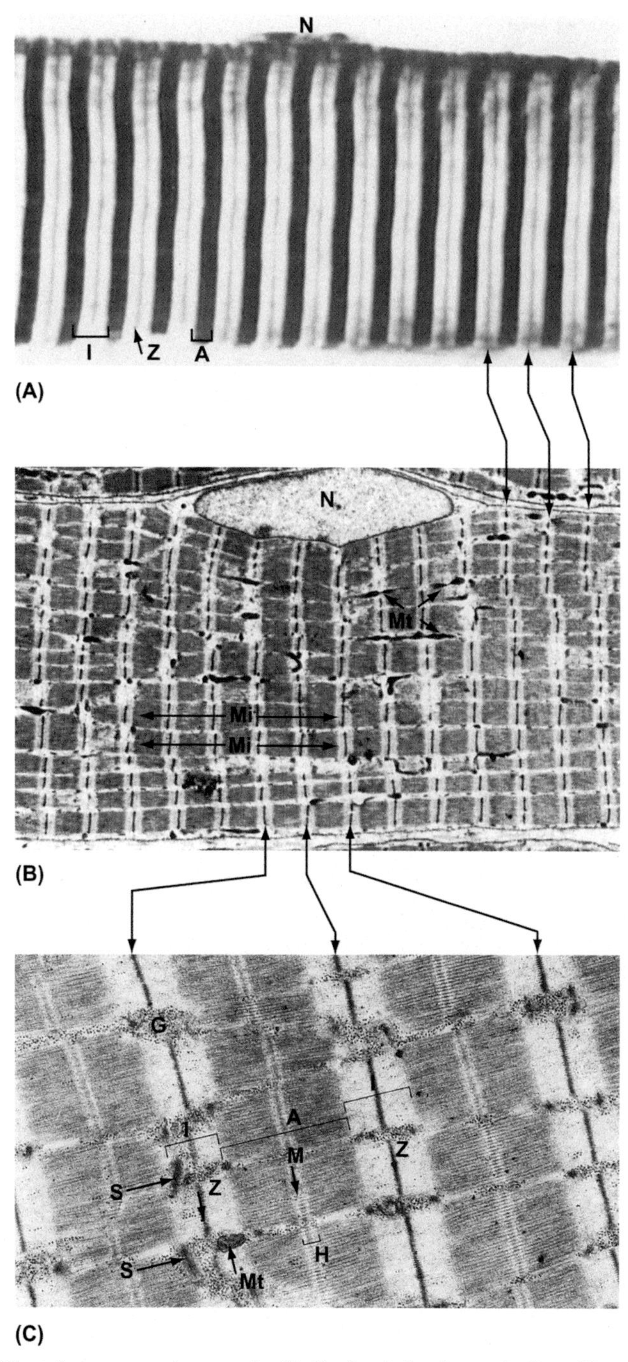

Figure 1.3 Light (A) and electron micrographs (B, C) of a skeletal muscle fiber, illustrating the banding patterns (A, H, I, M, Z) that reflect the high degree of organization of the contractile proteins. The two Z-bands are the boundaries of a sarcomere, the fundamental contractile unit of a muscle fiber. *G*, Glycogen deposits; *Mt*, mitochondria; *N*, nucleus; *S*, sarcoplasmic reticulum. *(From Young et al. (2006), with permission.)*

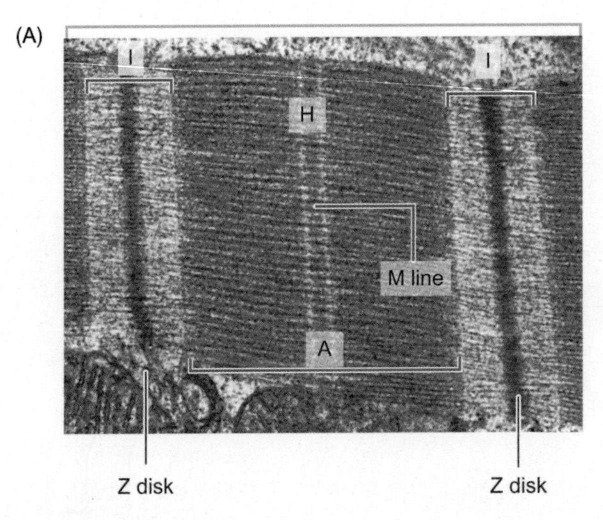

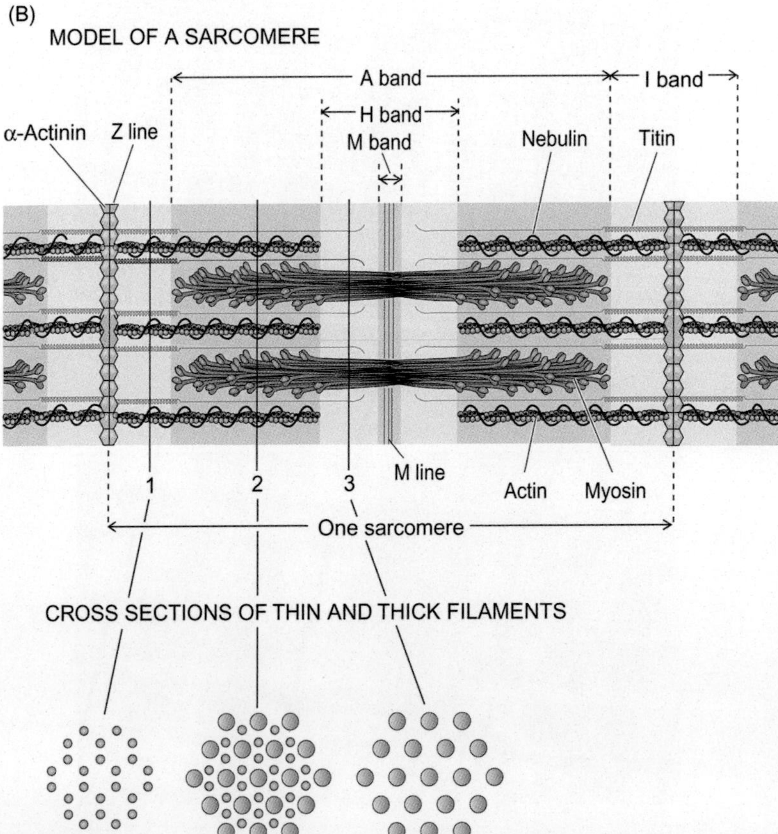

Figure 1.4 The structure of a sarcomere. (A) Electron micrograph. (B) Molecular model of a sarcomere. *((A) From Kierszenbaum and Tres (2012), with permission. (B) From Boron and Boulpaep (2012), with permission.)*

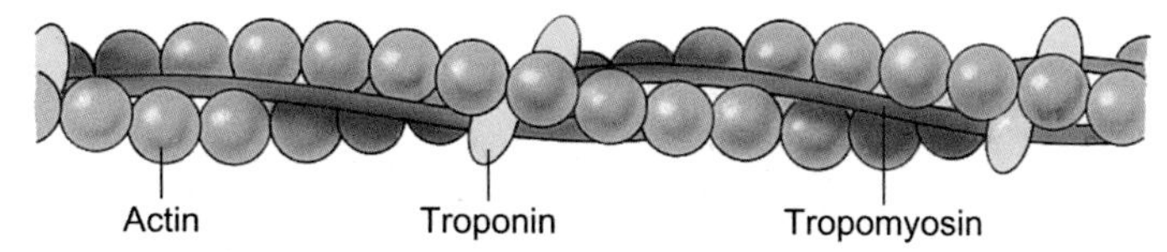

Figure 1.5 Structure of a thin filament, showing the relationship between tropomyosin and troponin to the actin chains. *(From Thibodeau and Patton (2007), with permission.)*

The Z band is a structural element composed of a variety of proteins, dominated by **α-actinin**, that help to connect adjacent sarcomeres, but equally important are **costameres**, molecular complexes that traverse the sarcolemma to connect the Z band proteins to the extracellular matrix. Two major complexes constitute a costamere. One is the **dystrophin–glycoprotein** complex (see Fig. 7.2); the other is the **integrin–vinculin–talin complex**. Both of these connect the terminal ends of the thin filaments to molecules that span the sarcolemma and connect to components of the extracellular matrix (endomysium) that surrounds individual muscle fibers. These costameral connections play important two-way roles in muscle function. In one direction, they are the means of transmitting the force generated from muscle contraction laterally to the connective tissue component of the muscle. This may actually exceed the longitudinal transmission of contractile force in a muscle. In the other direction, costameres represent a major means of transmitting external mechanical force into a muscle fiber and ultimately affecting the expression of genes guiding the formation of muscle proteins.

In addition to the filamentous contractile proteins, other intracellular structures are necessary for muscle fiber contraction. Surrounding each myofibril is a meshlike **sarcoplasmic reticulum** (Fig. 1.6), which plays a vital role in Ca^{++} storage and release. Interspersed among the myofibrils are varying numbers of **mitochondria** and granules of **glycogen**, both of which play important roles in providing energy for muscle contraction. Muscle fibers contain two main types of mitochondria—subsarcolemmal and intermyofibrillar. These mitochondria are interconnected to form a complex meshwork that serves to efficiently distribute energy production (see Fig. 7.10) throughout the muscle fiber.

A special adaptation that allows a muscle fiber to contract efficiently is regularly spaced **T-tubules**, which are inward extensions of the plasma membrane (sarcolemma) that are designed to facilitate the inward spread of the wave of electrical depolarization that initiates muscle fiber contraction. The presence of T-tubules is necessary because of the large diameter of a muscle fiber and the need to closely coordinate the contractile activity of all components within that muscle fiber.

The neuromuscular junction

A skeletal muscle is nonfunctional without innervation. The first step in the contractile process is the transmission of a signal to contract from a motor nerve to a skeletal muscle fiber. This occurs at the **neuromuscular junction**, a type of synapse between the

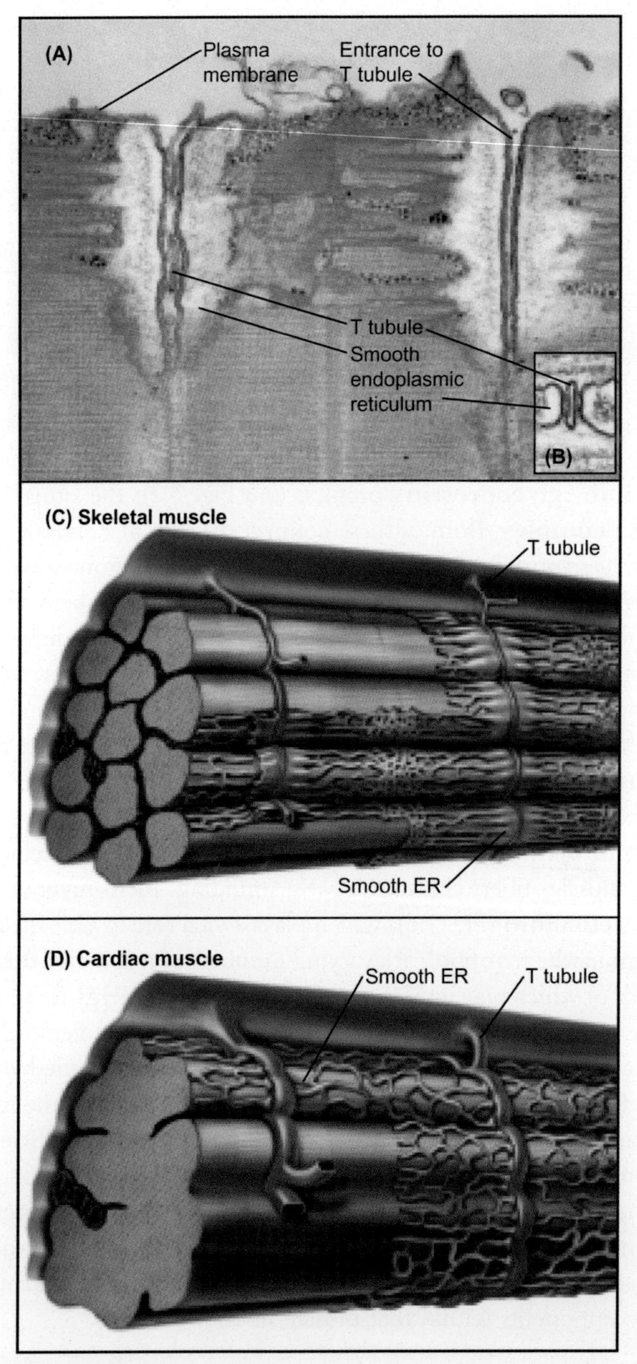

Figure 1.6 Membrane specializations of striated muscle fibers. (A) Electron micrograph of T-tubules penetrating into a muscle fiber from the plasma membrane (sarcolemma). (B) The close association between t-tubules and blind endings (cisternae) of the smooth endoplasmic reticulum (sarcoplasmic reticulum) is called a triad and is the site at which the contractile stimulus carried by the action potential on the plasma membrane is transferred to the interior of the muscle fiber. (C and D) Drawings of the three-dimensional arrangement of the T-tubules and sarcoplasmic reticulum in skeletal and cardiac muscle fibers. *(From Pollard and Earnshaw (2004), with permission.)*

terminal branch of a motor nerve fiber and a specialized region near the midpoint of a muscle fiber (Fig. 1.7).

The presynaptic nerve terminal, often called a **bouton**, does not differ greatly from other nerve terminals throughout the central nervous system. Numerous small

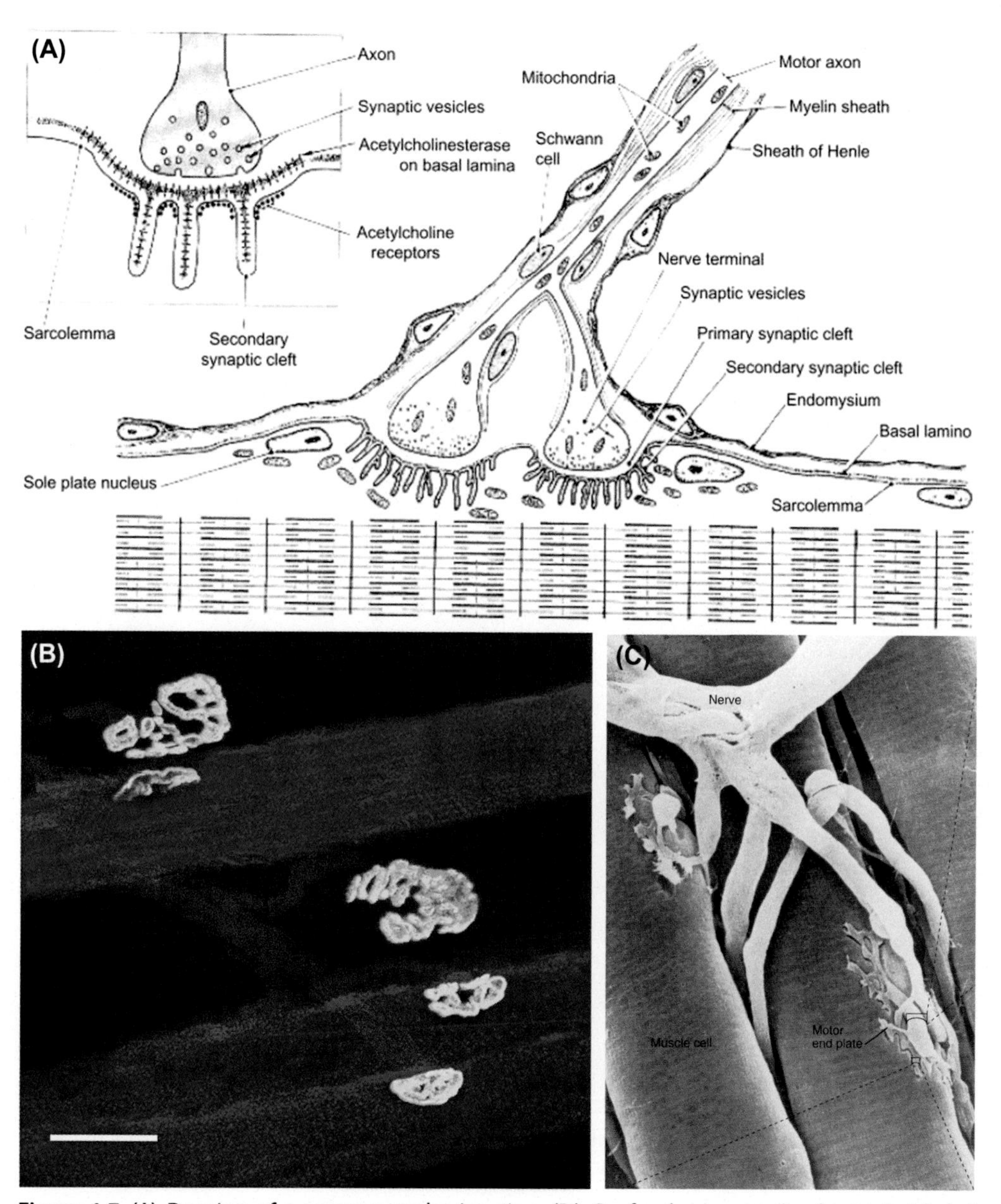

Figure 1.7 (A) Drawing of a neuromuscular junction. (B.) Confocal micrograph of a triple-labeled neuromuscular junction in the rat diaphragm. *Red*—Motor axons; *Blue*—muscle fibers; *Yellow/green*—motor endplates on the muscle fibers, labeled with α-bungarotoxin. (C) Scanning electron micrograph of a neuromuscular junction in frog muscle. *((B) From Mantilla, et al. Neuroscience 2007;146:181. (C) From Pollard and Earnshaw (2004), with permission.)*

(50–60 nm) vesicles containing the neurotransmitter **acetylcholine** are scattered throughout the bouton (Fig. 1.7). Some of these vesicles, each of which contains up to 10,000 molecules of acetylcholine, fuse with the plasma membrane of the nerve terminal. Upon simulation by an action potential passing down the axon of the motor nerve fiber, the vesicles release their contents of acetylcholine into the space between the nerve terminal and the muscle fiber.

The highly specialized region of the muscle fiber under the boutons of the motor nerve fiber differs from the rest of the muscle fiber and is often called the **motor endplate**. The sarcolemma of the motor endplate is thrown into **postsynaptic folds**, in which are embedded numerous **acetylcholine receptors**. The folding considerably increases the surface area in the endplate region and allows for a greater concentration of acetylcholine receptors. Muscle fiber nuclei (**sole plate nuclei**) in the vicinity of the motor endplate are specialized to produce the molecules required for maintaining the muscle (postsynaptic) side of the neuromuscular junction. Between the nerve terminal and the subjacent motor endplate lies a basal lamina, which is continuous with the basal lamina that surrounds the entire muscle fiber. Leaflets extending from this basal lamina extend into the spaces between neuromuscular folds (see Fig. 1.8). Embedded within the basal

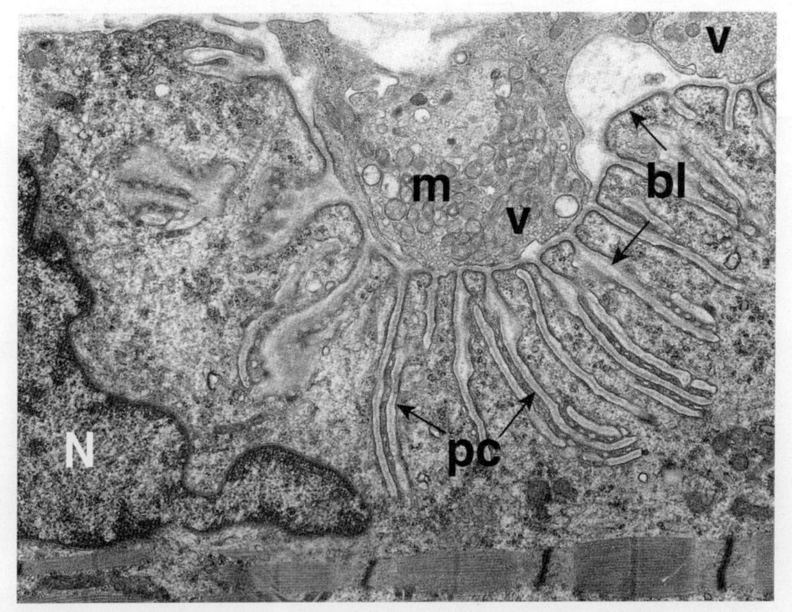

Figure 1.8 Electron micrograph of a human neuromuscular junction. *Bl*, basal lamina; *m*, mitochondria; *N*, nucleus; *pc*, postsynaptic clefts; *v*, vesicles. *(From Dubowitz (2013), with permission.)*

lamina are **acetylcholinesterase** molecules, which play an important part in modulating neuromuscular transmission.

Muscle fiber contraction

The contraction of a single muscle fiber involves almost all of the structural elements discussed above. A normal contraction begins with a signal from a motor nerve fiber to a muscle fiber and its propagation throughout the muscle fiber. The signal to contract is then translated into movement of the contractile proteins within each sarcomere. This produces a mechanical force and a shortening of the muscle fiber by about 20%. Then at the end of a contraction cycle, the muscle fiber must relax to its original resting length.

The Signal to Contract. The signal to contract begins with a wave of electrical depolarization (**action potential**) passing down the axon of a motor nerve fiber (Fig. 1.9). When the action potential reaches the bouton, voltage-gated Ca^{++} channels open, allowing Ca^{++} to enter the boutons. The increase in Ca^{++} concentration causes the synaptic vesicles that are located nearest to the surface fuse with the plasma membrane of the bouton. The fused vesicles then open and release their contents of acetylcholine into the extracellular space above the junctional folds. Active synaptic vesicles are concentrated opposite the receptor-rich ends of the synaptic folds, an arrangement that allows greater efficiency in neuromuscular transmission of a motor nerve impulse.

The acetylcholine released from the synaptic vesicles crosses the gap between the bouton and the junctional folds of the motor endplate and then activates the acetylcholine receptors that are concentrated at the tips of the folds. The acetylcholine receptor is an ion channel consisting of five subunits. When activated, the channel opens to the passage of Na^+ and K^+ from the extracellular fluid into the muscle fiber. This ion exchange increases the local membrane potential of the muscle fiber to the point where a propagated action potential begins to spread across the plasma membrane of the muscle fiber. Acetylcholinesterase, bound to the basal lamina material between the bouton and the motor endplate, breaks down the acetylcholine. Removal of the excess acetylcholine allows the neuromuscular junction to reset and be prepared to receive the next stimulus for the muscle fiber to contract.

Because skeletal muscle fibers are both very long and thick, it is important to have the stimulus for actual contraction of the muscle fiber arriving to all regions of the fiber as close to simultaneously as possible. The spread of the action potential along the length of the muscle fiber is a very rapid process, but penetration of this electrical stimulus into the interior requires a special structural adaptation, namely, the T-tubules, which

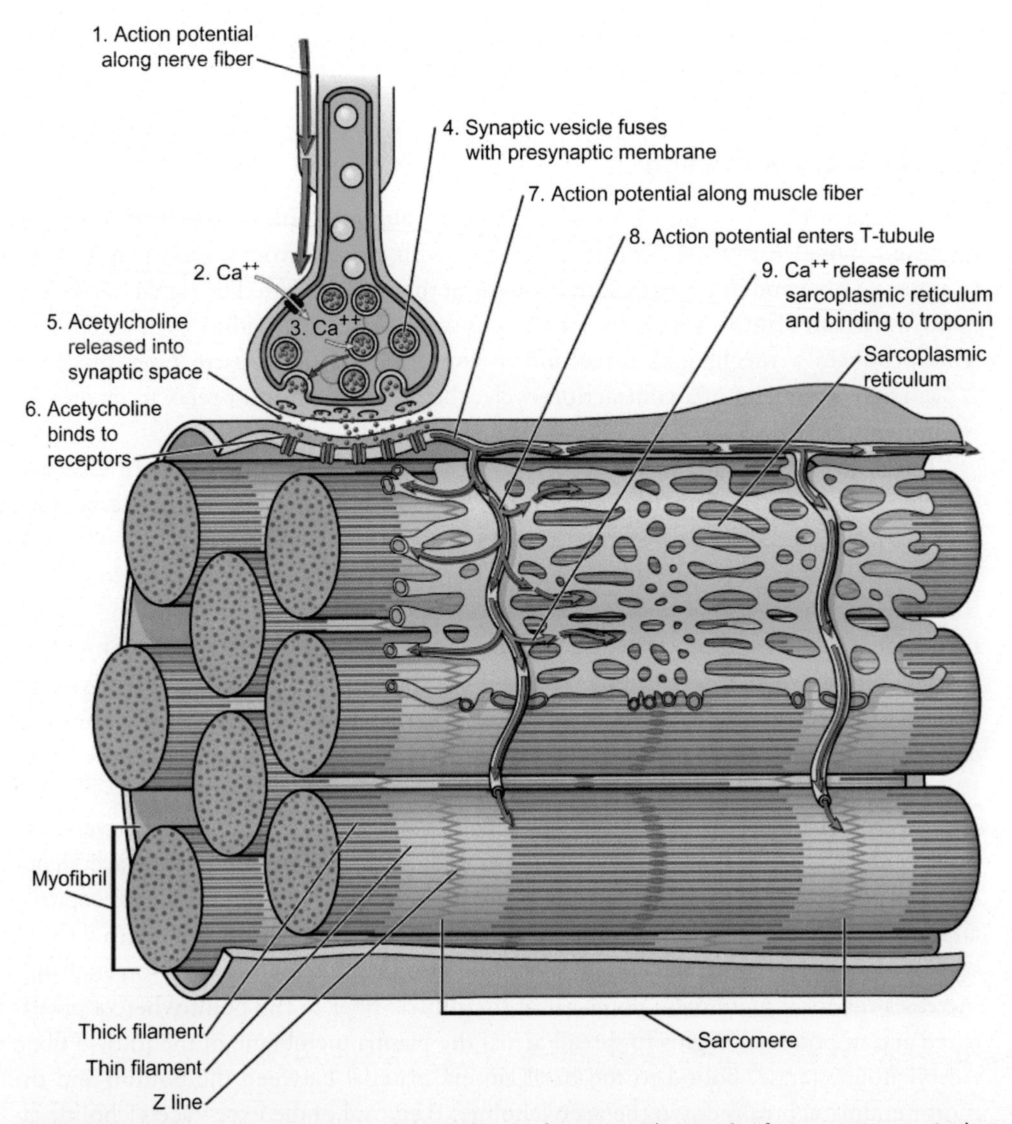

Figure 1.9 Schematic representation of the pathway of a contractile stimulus from a nerve terminal to the binding of Ca^{++} to troponin on the thin filaments. *(From Carlson (2019), with permission.)*

carry the action potential deep within the interior of the muscle fiber. T-tubules are found at every junction between the A and I bands of each sarcomere. At the myofibril level, T-tubules are sandwiched between expanded cisternae of the sarcoplasmic reticulum to form what are called **triads** (see Fig. 1.6 A,B). It is at the triads where the electrical stimulus to contract is converted to a chemical one in what is commonly called **excitation–contraction coupling**. Within the cisternae of the sarcoplasmic reticulum, the electrical stimulus opens Ca^{++}-release channels,[2] with the result that stored Ca^{++} ions are released from the sarcoplasmic reticulum into the cytoplasm of the muscle fiber, where they then bind to troponin and initiate the actual process of muscle contraction.

The Molecular Basis of Muscle Contraction. The molecular basis underlying muscle fiber contraction is the sliding of thin filaments past thick filaments, resulting in the shortening of the sarcomeres and consequently, the muscle fiber. This is accomplished by the formation of **cross-bridges** between the heads of the myosin molecules in the thick filaments and the actin subunits of the thin filaments. Actual motion (contraction) takes place through the coordinated bending of the myosin heads in a manner analogous to rowing a boat.

The formation of cross-bridges begins with the release of Ca^{++} from the sarcoplasmic reticulum. The released Ca^{++} binds to the troponin molecules, which are embedded within the thin filaments (see Fig. 1.5). Before Ca^{++}-binding, the tropomyosin strands block the myosin-binding sites on the actin molecules. Ca^{++}-binding to troponin causes a conformational change that results in the shifting of the tropomyosin strands deeper in the grooves between the two intertwined actin chains. This shift exposes the myosin-binding sites on the actin chains and sets the stage for binding between actin and the myosin heads of the thick filaments to form the cross-bridges.

The protruding myosin heads on the thick filaments contain two binding sites—one for actin and the other for ATP (adenosine triphosphate). Cross-bridge formation begins with the hydrolysis of the ATP on the myosin head to ADP and an inorganic phosphate group (P_i). As the myosin head, containing both ADP and P_i, begins to bind to actin, P_i is released. This causes a conformational change in the head of the myosin molecule, which allows actual binding to the actin-binding sites. At this point, the myosin heads protrude 90° from the long axis of the thick filament (Fig. 1.10).

Release of P_i from the myosin head and cocking of the head to 90° marks the beginning of the power stroke, during which the myosin head bends from 90° to 45°. The bending of many cross-bridges moves the thin filament 10−12 nm along the thick filament toward the center of the shortening sarcomere (Fig. 1.11). During this process, the ADP is released from the myosin head while the myosin cross-bridge remains attached to the actin of the thin filament. The cross-bridge remains intact until a new ATP becomes attached to the myosin head, thus releasing the myosin head from its actin-binding site.[3] When the cross-bridges are not in contact with their actin-binding sites, the muscle fiber is in a relaxed state. The contraction cycle continues when the

[2] The Ca^{++}-release channels are called **ryanodine receptors** because they can be blocked by a group of plant alkaloids, including ryanodine and caffeine.

[3] If the presence of ATP is exhausted, cross-bridges remain firmly in place. This causes a stiffening of the muscle and is the basis of **rigor mortis** that appears soon after death.

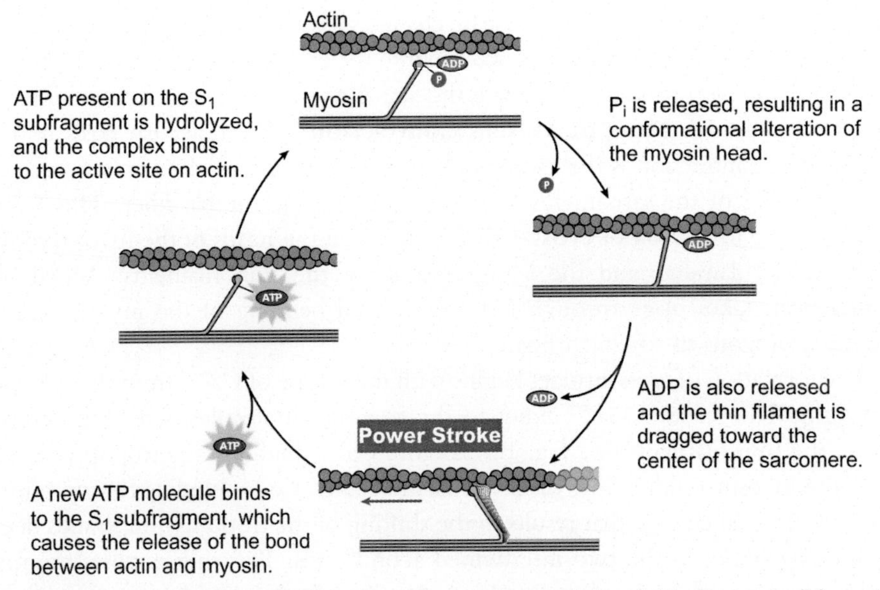

Figure 1.10 The role of ATP in propelling the power stroke as thick and thin filaments slide past one another during muscle fiber contraction. *(From Gartner and Hiatt (2011), with permission.)*

release of P_i from the ATP recocks the myosin head back to 90° in a location behind its previous site of attachment.

Muscle contraction continues as long as stimuli from the motor nerve cause action potentials to spread along the sarcolemma and into the T-tubules. Contraction stops upon cessation of the action potentials, but this requires the removal of Ca^{++} from the sarcoplasm. Ca^{++} removal is accomplished through the actions of a calcium pump that transports the calcium ions back into the sarcoplasmic reticulum. Within the sarcoplasmic reticulum, Ca^{++} is bound to a molecule called **calsequestrin**. Calcium pumps within the sarcolemma also remove some Ca^{++}, but the removal of too much by this means would deplete the muscle fiber of the Ca^{++} needed to initiate contraction.

Contractile properties of single muscle fibers and muscle fiber types

The shortening that occurs through the sliding of thin filaments past thick ones within a single sarcomere is multiplied many times to produce the contraction of a muscle fiber and the generation of force. At an even larger level, contractions of many muscle fibers produce contraction of a muscle as a whole. It is possible to measure the contractile properties of single muscle fibers. Knowledge of contraction at this level provides a good basis for understanding how entire muscles contract under various circumstances.

The simplest reaction of a muscle fiber to a single stimulus, whether acetylcholine release from a motor nerve terminal or a direct electrical stimulus in the laboratory, is

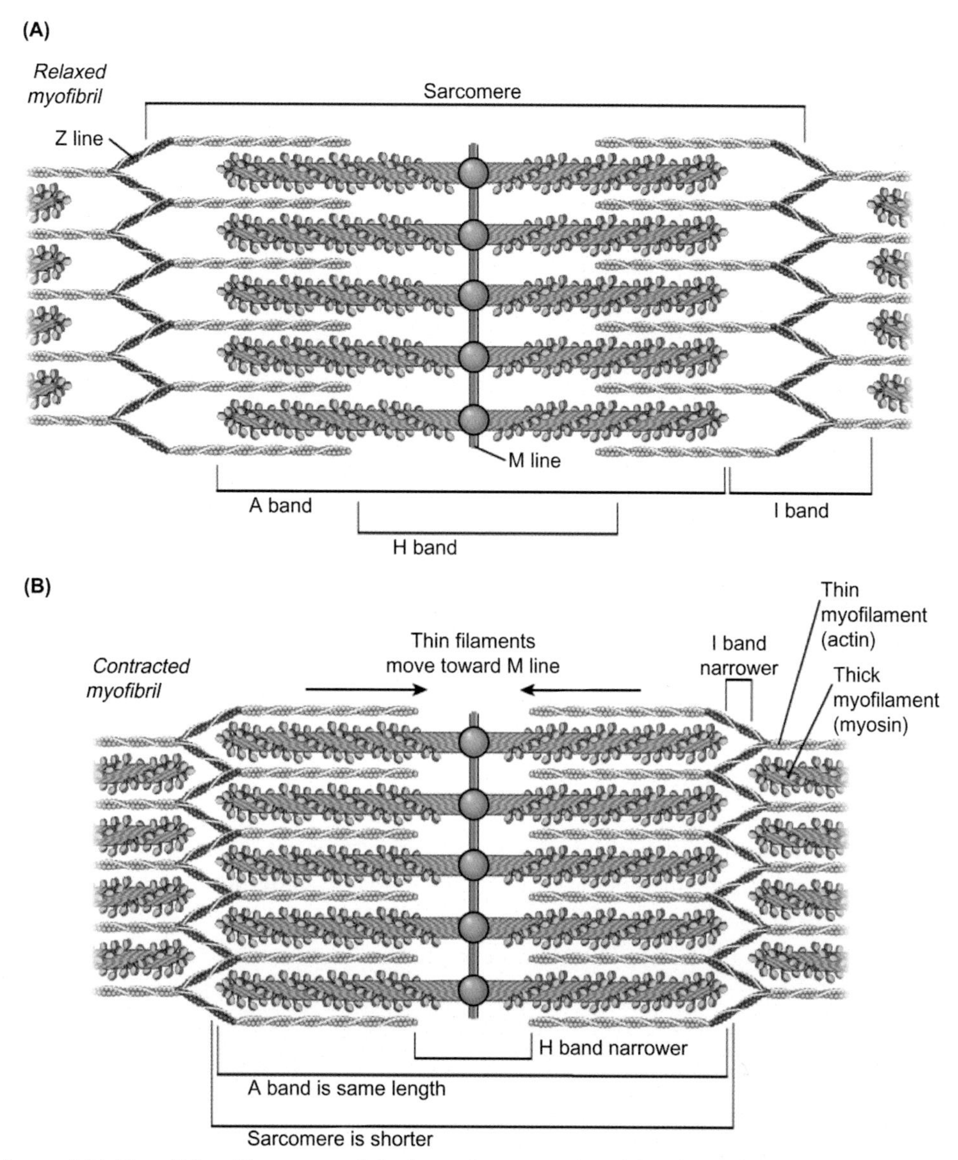

Figure 1.11 The sliding filament model of muscle contraction. (A) Relationship between thick and thin filaments in a relaxed muscle fiber. (B) The same, in a contracted muscle fiber. *(From Carlson (2019), with permission.)*

a **twitch** (Fig. 1.12). The curve representing twitch tension has three recognizable components. A brief latent period of several msec, in which no tension is generated after the initial stimulus, represents the time required for the action potential to pass along the surface of the muscle fiber and into the T-tubules, and then the time between the release of

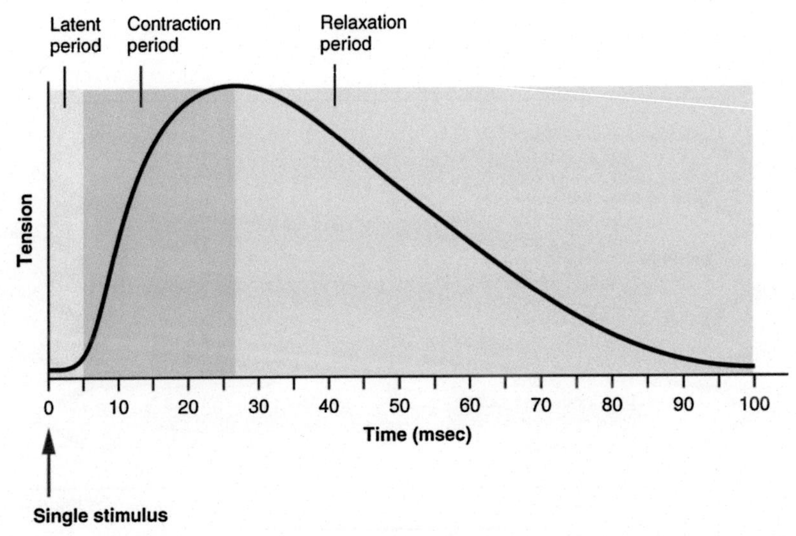

Figure 1.12 A breakdown of the basic elements of a twitch curve in a skeletal muscle fiber.

Ca^{++} from the sarcoplasmic reticulum and its effect on the tropomyosin of the thin filament. Once the Ca^{++} has freed up the myosin-binding sites on the actin chains and cross-bridges begin to form, tension begins to build up within the muscle fiber (contraction period), and as increasing numbers of cross-bridges are formed, it comes to a peak. This is followed by a longer relaxation phase during which tension gradually decreases as the number of cross-bridges becomes reduced to resting levels.

Not all muscle fibers have the same contractile properties. Muscle fibers have been divided into several major types according to their contractile properties and structural characteristics (Table 1.1). Type I (slow) muscle fibers contract most slowly to a single twitch stimulus (Fig. 1.13A). They also generate the least twitch tension. On the other hand, type I muscle fibers are much more resistant to fatigue than the other types. Three varieties of Type II muscle fibers (IIa, IIx, and IIb) exist. Type IIb muscle fibers are present in the commonly studied laboratory animals. Human muscle contains Type IIx, rather than IIb fibers. All Type II muscle fibers are more rapidly contracting than Type I fibers. Type IIb/IIx (fast, fatigable) is characterized by a fast twitch time and great twitch tension, but they become rapidly fatigued (see Fig. 1.13C). Type IIa (fast fatigue-resistant) is the slowest and also the most fatigue resistant of the Type II fibers. Differences between twitch times in both developing and adult muscles are largely the result of their containing different forms of myosins, in particular the ATPase activity associated with the particular myosins (see p. 59). Many different heavy chain myosins exist in striated muscle, both skeletal and cardiac, in different parts of the body and at different times of development (Table 1.2).

Table 1.1 Properties of the three major types of muscle fibers.

Characteristic	Type I	Type IIa	Type IIb or IIx
Contractile speed	Slow	Intermediate	Fast
Myosin ATPase activity	Low	High	High
Twitch tension	Low	Intermediate	High
Power	Low	Intermediate	High
Endurance	Highest	High	Low
Fatigability	Low	Intermediate	High
Recruitment	First (low load)	Intermediate	Last (highest load)
Fiber XS. area	Least	Intermediate	Greatest
Sarcoplasmic reticulum	Poorly developed	Intermediate	Highly developed
Z line	Thick	Intermediate	Thin
Neuromuscular junction area	Least	Intermediate	Greatest
Color	Red (myoglobin)	Reddish (myoglobin)	White
Mitochondria	Numerous	Most numerous	Least numerous
Metabolism	Oxidative	Oxidative	Glycolytic
Glycogen	Low	High	Highest
Glucose uptake	High	Intermediate	Low
Triglyceride content	Highest	High	Low
Capillary supply	High	Intermediate	Low

The vast majority of muscle contractions are not twitches. Rather they involve sustained contractions for various periods of time. In essence, when repeated stimuli occur so rapidly that the muscle fiber does not have time to relax before being called upon to contract again, the individual twitch contractions fuse and a **tetanic contraction** occurs. The characteristics of tetanic contractions differ among the three types of muscle fibers. Slow (Type I) muscle fibers require the lowest frequency of stimuli in order for tetanic fusion to occur, and when fusion has taken place, the contractile force is the lowest of the three types of muscle fibers. On the other end of the spectrum, Type IIb (fast fatigable) muscle fibers produce much more tetanic force, but they require a higher frequency of stimulation for tetanic fusion to occur. The characteristics of tetanic contractions of Type IIa (fast fatigue-resistant) muscle fibers lie between those of the other two muscle fiber types. These characteristics of the muscle fibers are matched by the firing frequencies of the associated motor nerve fibers, with those innervating slow muscle fibers discharging at a lower frequency than those innervating fast muscle fibers.

Human muscles typically contain a mix of the three types of muscle fibers, but in some animals, entire muscles are predominantly fast or slow. For example, the white meat of chickens is essentially fast muscle, whereas dark meat is slow, the color being

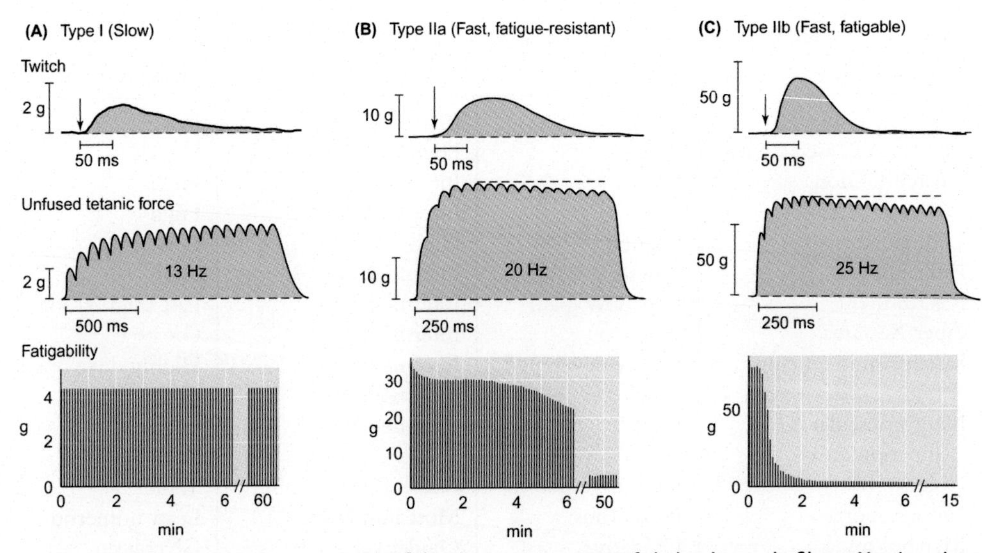

Figure 1.13 Contractile characteristics of the three main types of skeletal muscle fibers. X axis—time; Y axis—contractile force. (A) Slow (Type I); (B) fast, fatigue-resistant (Type IIa); (C) fast, fatigable (Type IIb). Top row—characteristics of a single twitch. Slow muscle fibers generate much less tension, but over longer period than do the other two types of fibers. Middle row—slow muscle fibers take longer to fuse individual twitches into a tetanic contraction, which is also relatively weak compared to that of fast muscle fibers. Bottom row—fatigability. Although relatively weak, slow muscles can contract at a maximum level for prolonged periods of time, whereas the strong contractions of fast muscles fade quickly because of fatigue. *(From Boron and Boulpaep (2012), with permission.)*

Table 1.2 Myosin heavy chain genes and proteins.

Gene	Protein	Location
MYH1	MyHC-2X	Type IIx fibers
MYH2	MyHC-2A	Type IIa fibers
MYH3	MyHC-emb	Embryonic fibers
MYH4	MyHC-2B	Type IIb fibers
MYH6	MyHC-α	Heart and jaw muscle fibers
MYH7	MyHC-β/slow	Heart and slow muscle (Type I) fibers; embryonic fibers
MYH7b	MyHC slow	Extraocular muscle fibers
MYH8	MtHC-neo	Fetal muscle fibers
MYH13	MyHC-EO	Extraocular muscle fibers
MYH15	MyHC-15	Extraocular ocular muscle fibers
MYH16	MyHC-M	Jaw muscle fibers

due to the high content of the oxygen-binding protein **myoglobin** within the muscle fibers. The role of the various muscle fiber types in relation to the organization and function of entire muscles will be covered later in the book.

Motor units

Although the muscle fiber, a single multinucleated cell, may be considered the fundamental structural component of a muscle, the fundamental functional component is the **motor unit**, which consists of a single branching motor nerve fiber and all the muscle fibers to which it is attached (Fig. 1.14). When a motor nerve is stimulated, all of the muscle fibers comprising that motor unit contract simultaneously. All muscle fibers within a

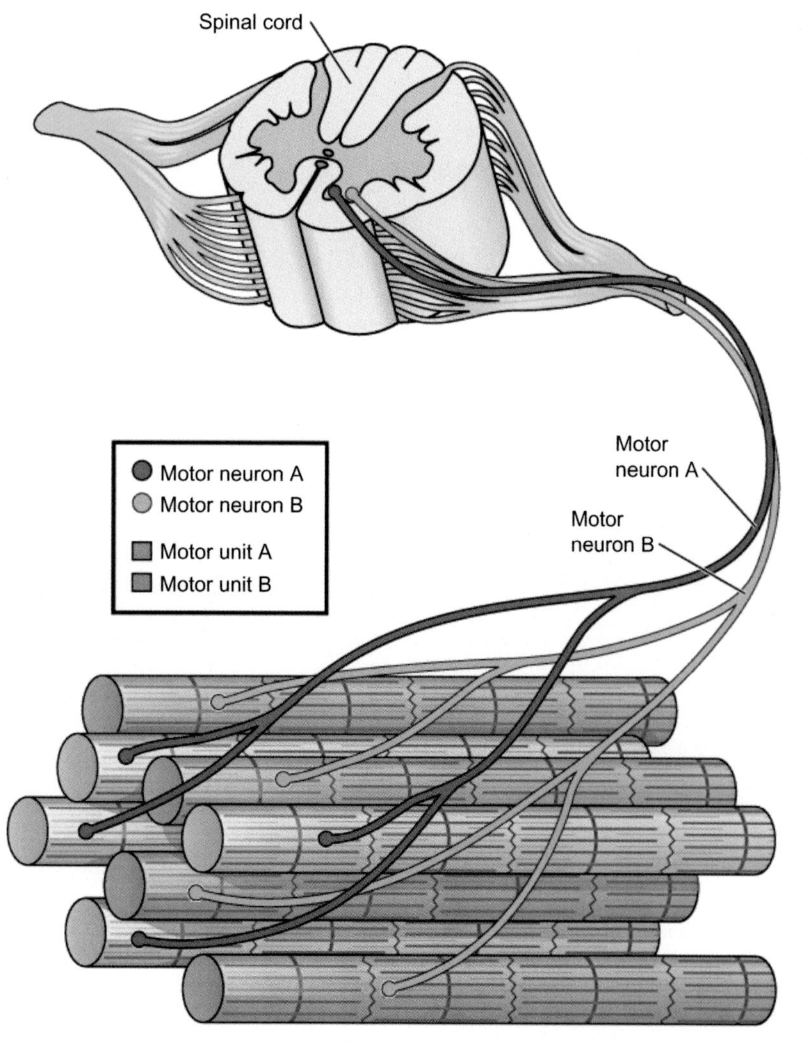

Figure 1.14 Diagrammatic representation of two motor units in a skeletal muscle. *(From Carlson (2019), with permission.)*

motor unit are of the same type. Even though the muscle fibers within a given motor unit have the same characteristics, a single human muscle usually contains motor units of different functional types, which are served by different types of motor nerves.

Motor units are important in providing graded contractile responses within a muscle, and their size corresponds to the size and functions of the muscle. Although an average motor unit may consist of 100 muscle fibers attached to a single nerve fiber, they range in size from as few as four muscle fibers in the small extraocular muscles to as many as 1000 in the larger muscles of the thigh. The muscle fibers of a motor unit are not all collected in a single area, but rather they are spread around throughout a broad region of a muscle. Conversely, a given region of a muscle will contain muscle fibers from many different motor units. As a result, fast and slow muscle fibers are interspersed in what is often called a checkerboard pattern (Fig. 1.15). Such a distribution allows smoother overall contractions of a muscle, and it reduces the chances of local stress that could happen if all the muscle fibers in a small area contract at the same time.

The sensory apparatus of a muscle

The normal function of a muscle requires sensory feedback (**proprioception**) from the muscle to the brain or spinal cord. This is accomplished through two quite different sensory receptors: one located within the belly of the muscle itself and the other within the

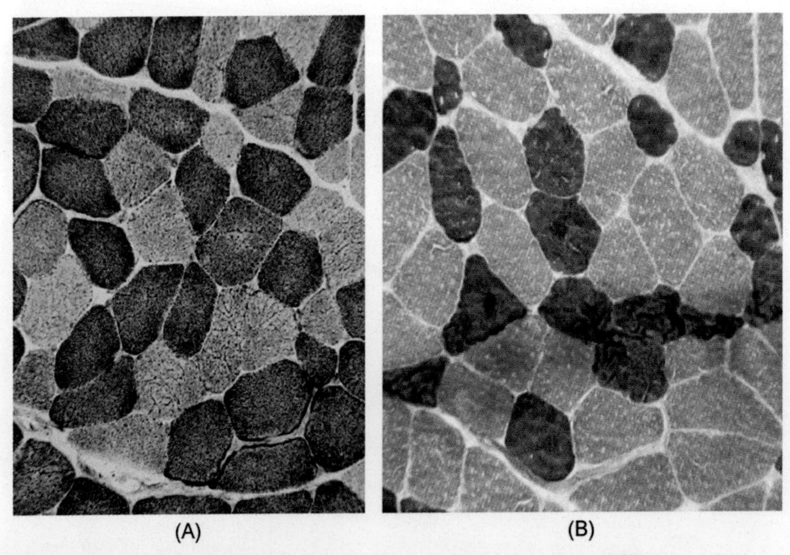

Figure 1.15 Histochemical preparations of skeletal muscle, showing differences between Types I (slow) and II (fast) muscle fibers. (A) Type I fibers have higher concentrations of oxidative enzymes than Type II fibers. (B) In this preparation, Type II fibers stain more strongly for myofibrillar ATPase than do Type I fibers. *(From Nolte (2009), with permission.)*

tendon. Near the myotendinous junction within the tendons at each end of a muscle, **Golgi tendon organs** are encapsulated structures consisting of sensory nerve fibers enwrapping strands of collagen that are continuous with 10–20 motor units within the muscle (Fig. 1.16). The capsule consists of fibrous connective tissue. When the muscle contracts, the nerve endings are compressed and sense the degree of tension within the tendon as a measure of the amount of force generated by the muscle. This activates stretch-sensitive cation (positively charged ions) channels and depolarizes the nerve endings, which then send an action potential to the spinal cord for further processing by the central nervous system. Golgi tendon organs are able to sense the force generated by the muscle throughout its entire physiological range.

Muscle spindles (see Fig. 1.16) function in an entirely different manner. They measure the length and rate of stretch of a muscle. Situated in variable locations throughout the belly of the muscle, muscle spindles differ from Golgi tendon organs by functioning in parallel with other muscle fibers, rather than in series, as is the case for tendon organs. Structurally, muscle spindles are quite complex. A typical muscle spindle contains 3–12 small **intrafusal muscle fibers**, which are of two varieties—**nuclear bag** or **nuclear chain**. With respect to muscle spindles, ordinary muscle fibers are called **extrafusal**

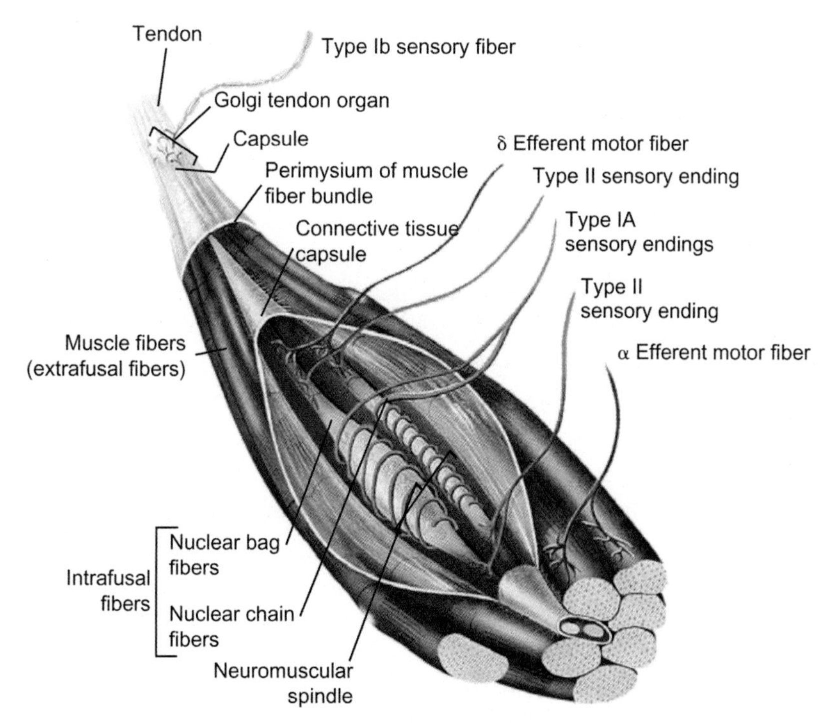

Figure 1.16 Diagrammatic representation of a muscle spindle and Golgi tendon organ in a skeletal muscle. *(From Thibodeau and Patton (2007), with permission.)*

muscle fibers. Nuclear bag fibers contain a clump of nuclei in the central region of the fiber, whereas the nuclei of nuclear chain fibers are spread out in a single longitudinal row within the central region of the fiber. Both types of intrafusal muscle fibers contain arrays of contractile proteins at either end of the fiber, but not in the middle, nucleated region. Two types of sensory nerve fibers, as well as motor axons, innervate the intrafusal fibers. Type Ia sensory nerve fibers wrap around the central regions of both bag and chain muscle fibers. In contrast, endings of Type II sensory nerve fibers connect with both ends of the intrafusal fibers. When the muscle is stretched or is maintained in a stretched condition, these sensory nerve fibers increase their signaling to the central nervous system.

Nuclear chain and one type of nuclear bag fibers are static sensors, whereas another type of nuclear bag fiber serves as a dynamic sensor. Muscle spindles function well only if the central regions of the intrafusal fibers are kept at a certain degree of tension by the contractile ends of these fibers regardless of the state of contraction of the muscle as a whole. When the central region is stretched, stretch-sensitive ion channels in the Type Ia sensory nerve fibers innervating that region are activated and Na^+ rushes in, stimulating an action potential leading to the spinal cord. This, in turn, sends the signal toward centers in the central nervous system, where further processing occurs.

The contractile portions of the intrafusal fibers and their motor innervation play an important role in their overall function. Upon contraction of the extrafusal fibers in response to stimulation by α-motor neurons, the intrafusal fibers within the muscle spindle would go slack, but acting in parallel with the extrafusal fibers, γ-motor neurons stimulate the intrafusal fibers to contract. This restores an appropriate degree of tension to the intrafusal fibers and allows the sensory nerves of the spindle to modulate the state of contraction of the entire muscle.

For any given muscle, the density of muscle spindles corresponds to its function. A muscle with very sensitive functions has a higher density of spindles than one involved in gross movements. Whereas the gluteus maximus muscle contains only one spindle per gram of tissue, extraocular muscles have as many as 36 spindles per gram.

Other proprioceptive sensors are actually situated outside muscles themselves. Deep **Pacinian** corpuscles register pressure, whereas within fascia, **Ruffini endings** detect tension and shear. Ligaments associated with joints also contain as many as four types of mechanoreceptors. These receptors all send signals to the brain, which is able to integrate the sensory information that it receives and translate it into smooth movements of the muscles.

Contraction of an entire muscle

The contraction of an entire muscle is a highly coordinated event that requires the participation of motor units to a degree that corresponds to the need to generate a specific amount of force. How many and what kinds of motor units is a function of the force required. Although volition plays a significant role as a factor in muscle contraction, sensory input from both the belly of the muscle and its tendons is also a major contributor to the overall process of muscle contraction.

At the whole muscle level, contraction is a graded, rather than an all-or-nothing event. A minor movement involves a considerably different strategy from contracting against a maximal load. To counteract a very minor load, such as the need to maintain a specific posture, the participation of only a few motor units is required. These are small ones composed of slow muscle fibers, which produce relatively weak, but sustained contractions. As the load on a muscle increases, larger and more powerful motor units, composed of Type IIa muscle fibers, are called into play. When the muscle needs to contract against a maximum load, Type IIb (or IIx) motor units, with the largest and most powerful muscle fibers are activated. Because these Type IIb and Type IIx motor units are composed of muscle fibers with a glycolytic metabolism, a muscle is unable to sustain a maximal contraction for any length of time.

Architecturally, many muscles contain a higher concentration of slow motor units near its concave edge or closest to the bone. Correspondingly, the Type IIb motor units tend to be situated on the side away from the closest bone. Overall, in humans the upper limbs contain a higher proportion of fast muscle fibers than do the lower limbs. During a sustained muscle contraction, motor units are recruited at staggered intervals, which results in a smooth contraction of the muscle. During a less than maximal contraction, a given level of force can be maintained by either recruiting different motor units or by changing the firing rate of those motor units that have been activated. As a muscle relaxes after a sustained contraction, motor units are decommissioned in reverse order, with large fatigable ones becoming quiescent first and followed by successively smaller ones.

The role of internal architecture in muscle function

In addition to the intrinsic contractile properties of individual muscle fibers, their architectural arrangement within a muscle is an important determinant of how a muscle contracts. This is well illustrated by examining the flexor and extensor muscles of the human thigh. The flexor muscles (the hamstrings) are long and narrow, and within them, the muscle fibers are largely oriented parallel to the long axis of the muscle. On the other hand, the extensor muscles (the quadriceps [quads]) are bulkier, and internally the muscle fibers emerge at an angle from a central tendinous area much like the barbs in a feather. These architectural differences play an important role in how the muscles function. The contracting hamstring muscles shorten more, but are weaker, whereas the quads produce much more force, but less **excursion** (the range of extension or shortening) than the hamstrings. At a practical level, observe what is required for running or climbing stairs. The quads must elevate the weight of the body; on the other hand, the hamstrings lift the lower leg and foot.

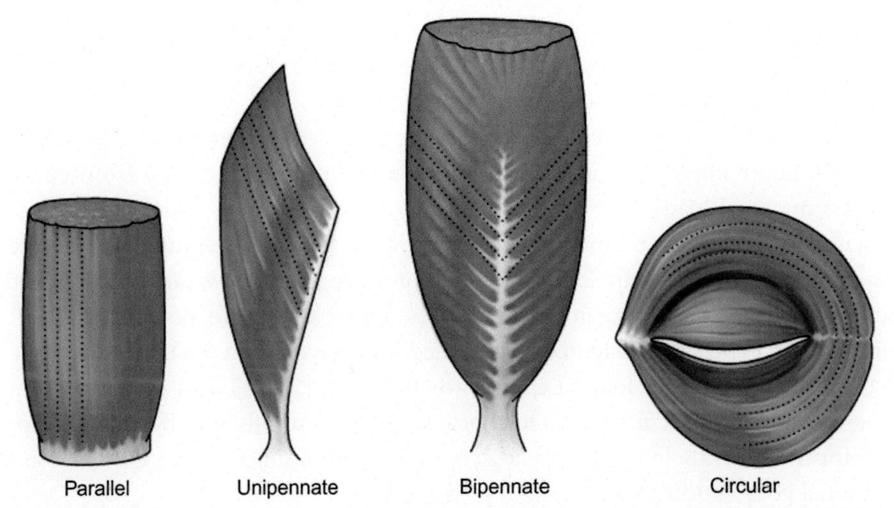

Figure 1.17 Fundamental types of internal architecture in skeletal muscles. *(From Carlson (2019), with permission.)*

Muscles have a variety of internal architectures that depend upon their location and function (Fig. 1.17). The simplest is parallel arrangement of the muscle fibers. In a long strap muscle, the length of the muscle exceeds the length of the individual muscle fibers, which are arranged in series. In muscles of this type, there are often several sites along the muscle where nerves and blood vessels enter and leave as a means of coordinating contractile activity throughout the muscle. Other muscles have muscle fibers coming off a central connective tissue core in either a **unipennate** or a **bipennate** fashion (see Fig. 1.17). In a few areas of the body, especially around orifices or the eyes, muscles are circularly arranged. The tongue has a very complex three-dimensional arrangement of intrinsic muscle fibers, which allows the very complex movements, including extension, of that structure. Such a three-dimensional architecture is also seen in the trunk of elephants and the tentacles of an octopus or a squid, both of which are capable of equally complex movements.

The overall contractile force of a muscle is a function of its **physiological cross-sectional area**, not its **anatomical cross-sectional area** (see Fig. 1.18). The anatomical cross-sectional area measures the greatest cross-sectional area of the muscle. In contrast, the physiological cross-sectional area is measured across the long axis of the muscle fibers, and not necessarily the whole muscle. In a parallel-fibered muscle, the anatomical and physiological cross-sectional areas are essentially the same. In contrast, a bipennate muscle, such as the rectus femoris muscle in the thigh, has a much larger physiological than anatomical cross-sectional area because of the angle at which the muscle fibers come off its central tendon. The soleus muscle of the calf has a high degree of pennation but also contains a large percentage of slow muscle fibers. This allows the

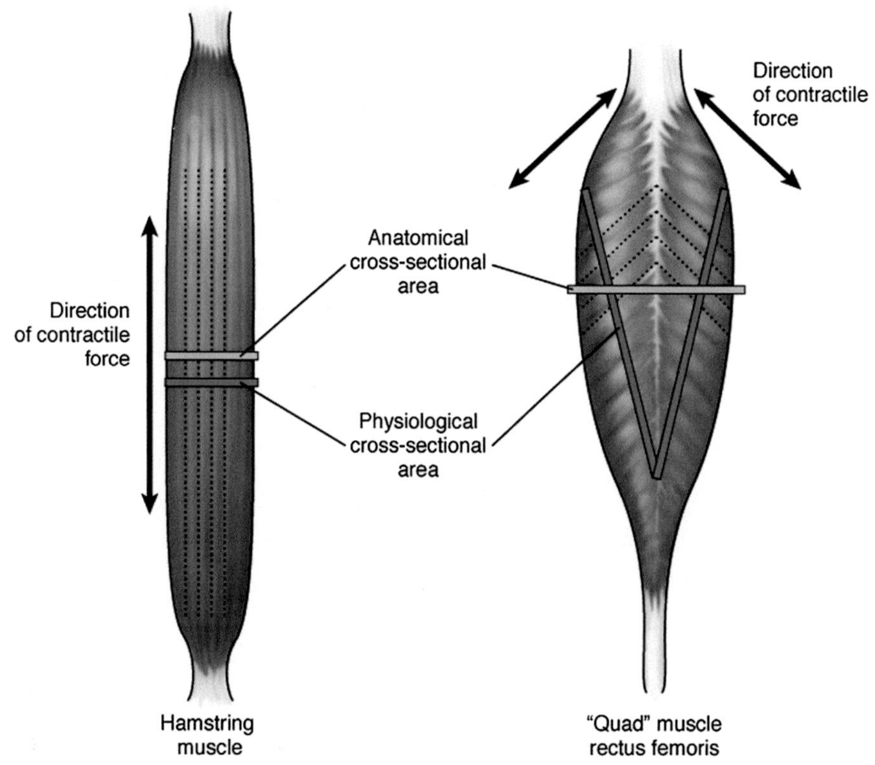

Figure 1.18 Functional architecture of flexor (hamstring) versus extensor (quadriceps) muscles in the human thigh. *(From Carlson (2019), with permission.)*

soleus to operate as both a major postural muscle and one that can generate a large amount of force when needed. As noted above, there are trade-offs to these arrangements. A bipennate muscle is capable of powerful contractions at the expense of a more limited excursion, whereas a parallel-fibered muscle exerts less force, but at a longer distance. The importance of cross-sectional area is important in training (see Chapter 4) because strength is directly related to the "size" of the muscle.

The role of connective tissue in whole muscle contraction

The role of connective tissue in whole muscle contraction is often greatly underestimated. It is the connective tissue surrounding the muscle fibers and the muscle itself that translates the contractile force developed by the muscle fibers into actions that move or support parts of the body. This transmission of force begins at the level of the muscle fiber. The ends of a muscle fiber connect directly with the collagen fibers of tendons through specialized **myotendinous junctions**. These will be dealt with in the next section. In addition, the lateral borders of muscle fibers (sarcolemma) are attached to the surrounding endomysium (see Fig. 1.1). This connection involves a series of molecules beginning within the muscle fiber at the level of the actin filaments, before passing through the sarcolemma and then binding with **laminin**, a major component of the basal lamina that closely invests the muscle fiber (Fig. 1.19).

The major molecular link between the contractile apparatus (thin filaments) the sarcolemma is the very large subsarcolemmal protein **dystrophin.**[4] At the sarcolemma, dystrophin connects to an aggregation of sarcolemmal linker proteins (see Fig. 1.19), which outside the muscle fiber connect to laminin and other components of the basal lamina. Other links connect laminin to the endomysial connective tissue.

The wispy endomysium, which surrounds individual muscle fibers, smoothly blends in with the tougher and more prominent perimysium that surrounds bundles of muscle fibers. The perimysium and the muscle fibers that it encircles constitute a **muscle fascicle**. Toward the ends of a muscle, the perimysial connective tissue converges with the dense collagenous connective tissue of the tendon. Added to this is the connective tissue of the epimysium, which covers the entire muscle. The tapered ends of a muscle allow the contractile forces produced by the muscle fibers to converge upon the relatively small cross-sectional area of the tendon.

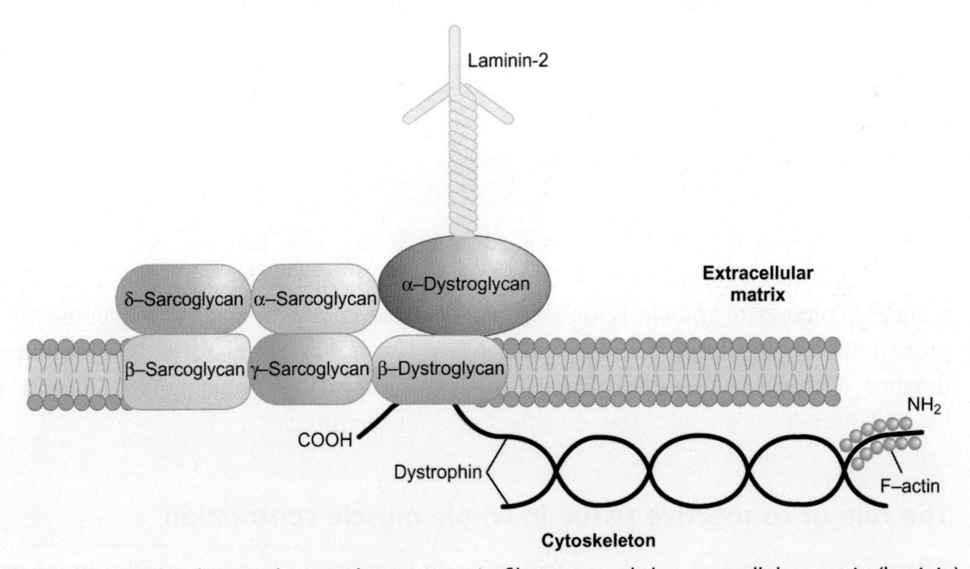

Figure 1.19 Molecular attachments between actin filaments and the extracellular matrix (laminin). *(From Jorde et al. (2010), with permission.)*

[4] Dystrophin is the protein that is defective in Duchenne muscular dystrophy, a sex-linked genetic disease affecting young boys. See Chapter 7 for further details.

The connective tissue of a muscle exhibits considerably different mechanical properties from those of the muscle fibers themselves. Whereas muscle fibers are both contractile and elastic, connective tissue is elastic. Connective tissue lengthens in response to prolonged stretch and tends to shorten if not stretched on occasion. This property is the reason why stretching is important in preventing hamstring injuries in runners.

Tendons and their connection to muscle and bone

Grossly, a tendon is the glistening white structure that connects the end of a muscle to a bone. In contrast to common representations of tendons, the muscular end of a tendon often spreads out along the surface of a considerable portion of the length of the muscle like a partially opened fan, thus providing a broad surface for the attachment of the ends of muscle fibers. The bulk of a tendon has a relatively simple structure—long bundles of tightly packed type I collagen fibers, with a sparse vascular supply and a small number of fibroblasts, called **tenocytes**, interspersed among the bundles of collagen. Golgi tendon organs have already been described above. Despite the apparent simplicity of the main part of a tendon, the structure of either end is relatively complex. The main function of the ends of tendons is to maintain a firm connection with either the muscle fibers or the bone to which the tendon is attached and to resist rupture at high tensile forces.

In some locations, for example, the abdominal wall and the ribs, muscles do not end as discrete tendons. Rather, they connect with broad stretches of fibrous connective tissue in the form of an **aponeurosis**. Such an arrangement distributes the contractile force of the muscle over a broad area instead of concentrating it on a single small focus.

The myotendinous junction

The myotendinous junction is a highly specialized region in which the sarcolemma extends from the end of a muscle fiber as a group of many fingerlike projections (Fig. 1.20). As is the case in many parts of the body, such projections increase the surface area for a variety of functions—in this case a mechanical one. Within these projections are actin filaments that project from the last Z-bands and continue to the ends of the projections. At the tips of the fingerlike projections are two groups of proteins (the same ones found in costameres) that connect the intracellular actin filaments to laminin molecules in the basal lamina outside the muscle fiber. One of these molecular groups consists of dystrophin and its associated membrane proteins (the dystrophin–glycoprotein complex). The other includes members of the **integrin** family of molecules. Integrins act as important connecters between cell membranes and the extracellular matrix. On the tendon side, laminin and other basal lamina molecules connect principally with **type I collagen** and another matrix protein called **tenascin**, both of which are produced by the tenocytes resident within the tendons. The strength of these

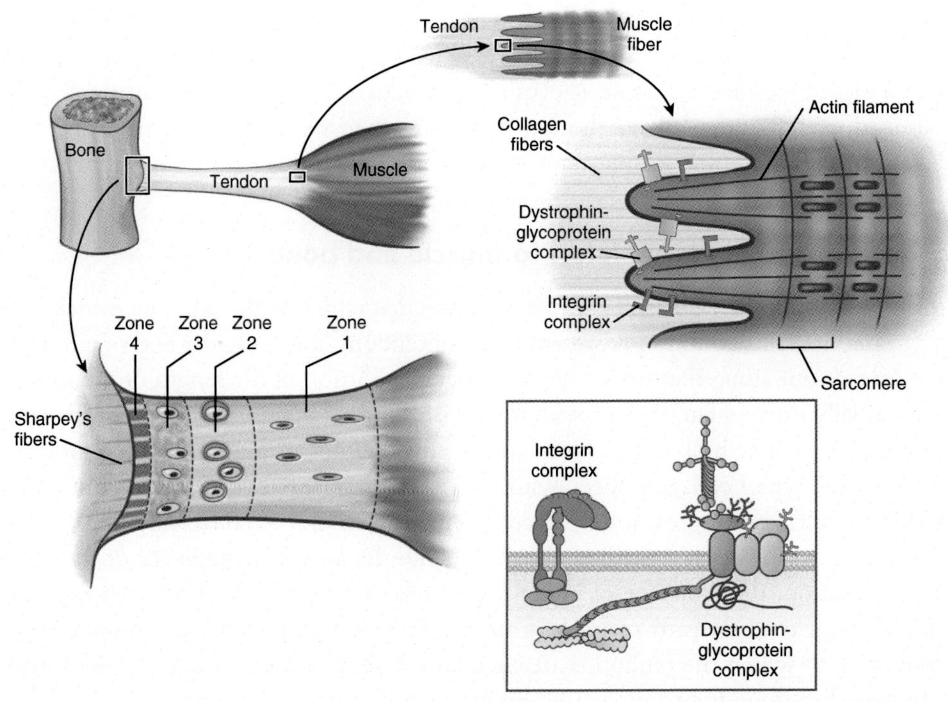

Figure 1.20 Attachments of a tendon to bone (enthesis) and a muscle (myotendinous junction). See the text for an explanation of zones 1–4.

connections is such that almost never does the rupture of a muscle occur in this region. Rather, it can occur toward the belly of the muscle or within the tendon itself. Through the individual connections of muscle fibers with the tendon material at the fan-shaped muscular ends of the tendons, the force generated by the contracting muscle fibers begins to become concentrated into the relatively small cross-sectional area of the tendon itself.

The enthesis—the connection between tendon and bone

In order to effect movement of body parts, a muscle and its tendon must also be firmly connected to a skeletal element. This connection is called an **enthesis**, and like the myotendinous junction, the enthetic attachment must be strong enough to resist rupturing when the muscle is contracting with maximal force.

A typical enthesis consists of four structural zones (see Fig. 1.20). The first is the body of the tendon itself, with massive parallel bundles of Type I collagen fibers and a relatively small number of tenocytes, which are relatively quiescent in a mature tendon. As the tendon approaches the bone, the tenocytes surround themselves with small amounts of a cartilaginous matrix, a common adaptation to a poorly vascularized environment.

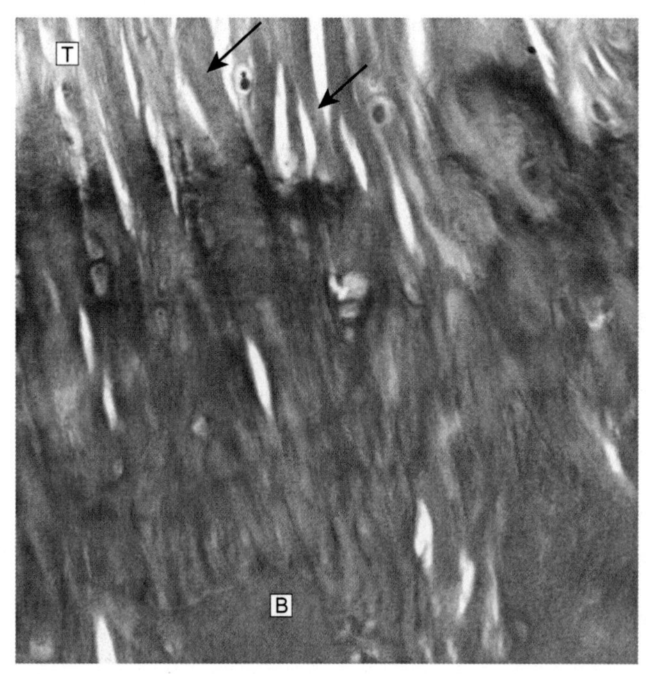

Figure 1.21 The site of a tendon insertion in to a bone. Collagen fibers (*arrows*) from the tendon (T) insert into the bone (B), at which point they are known as Sharpey's fibers. *(From Stevens and Lowe (2005), with permission.)*

This form of connective tissue is known as **fibrocartilage**. The third zone, directly opposing the bone, consists of calcified fibrocartilage, a tissue in which the cartilage matrix has hardened due to the deposition of calcium salts, in particular calcium phosphate. The fourth zone is the bone interface with the end of the tendon. What makes this zone important is the presence of **Sharpey's fibers**, tough strands of collagen fibers extending from the tendon and firmly embedded in the bone matrix (Fig. 1.21).

Muscle contractions and whole-body movements

Most muscles are designed to move a body part across a joint. Especially in the limbs, muscles are sometimes classified as one-joint or two-joint muscles. Often, such muscles are in pairs and with similar functions. In the thigh, for example, the long head of the biceps femoris muscle, which is a major flexor of the leg, is a two-joint muscle, whereas the underlying short head of that muscle is a one-joint muscle (Fig. 1.22). In the arm, the biceps brachii and brachialis muscles are similarly organized. Two-joint muscles are long and superficial. They are powerful and move the joints that they cover in a variety of ways. In contrast, their one-joint counterparts are located close to the bone and typically serve more postural functions.

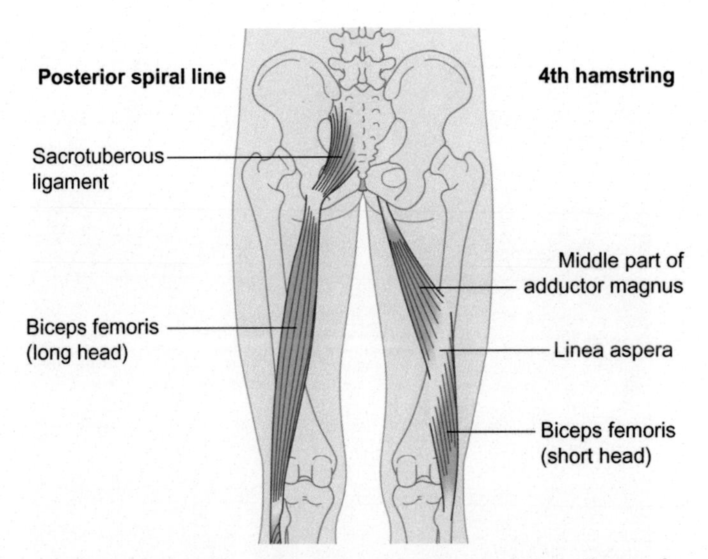

Figure 1.22 Two-joint (e.g., biceps femoris, long head) versus single-joint (biceps femoris short head and adductor magnus) muscles. Two-joint muscles play an important power function, whereas single-joint muscles are more important in adjusting posture. *(From Myers (2014), with permission.)*

Most muscles function as levers, and where they attach to a bone plays an important role in how they function. As applied to muscle, a lever system has three components—a fulcrum, a load, and a muscle whose contraction counteracts that load (Fig. 1.23). A good starting example is the biceps brachii muscle, which attaches to the proximal end of the ulna. In this case, the fulcrum is the elbow joint, and the load is the hand or a load grasped by the hand. This arrangement represents a classic Type III lever system, in which the muscle and the load are on the same side of the fulcrum, and the load is further from the fulcrum than the muscle attachment. Without getting into the trigonometric basis, where the biceps attaches to the ulna plays a significant role in the amount of load that the muscle can counteract. The farther distally the biceps inserts into the ulna, the greater its mechanical advantage, but this advantage would be counteracted by a greater limitation to the extension of the forearm.

A similar situation is seen in a Type II lever (see Fig. 1.23), but in this case the muscle attachment is farther from the fulcrum than the load. This type of arrangement allows more efficient use of the contractile power of the muscle in working against a load, but this mechanical advantage is gained at the expense of a reduction in distance moved. A good example of this is the attachment of the Achilles tendon to the heel (calcaneous bone), which allows the muscles of the calf to lift considerable weight over a short distance.

The third class of lever (Type I) functions is in the manner of a see-saw, where the muscle and its load are on opposite sides of the fulcrum. This type of arrangement is

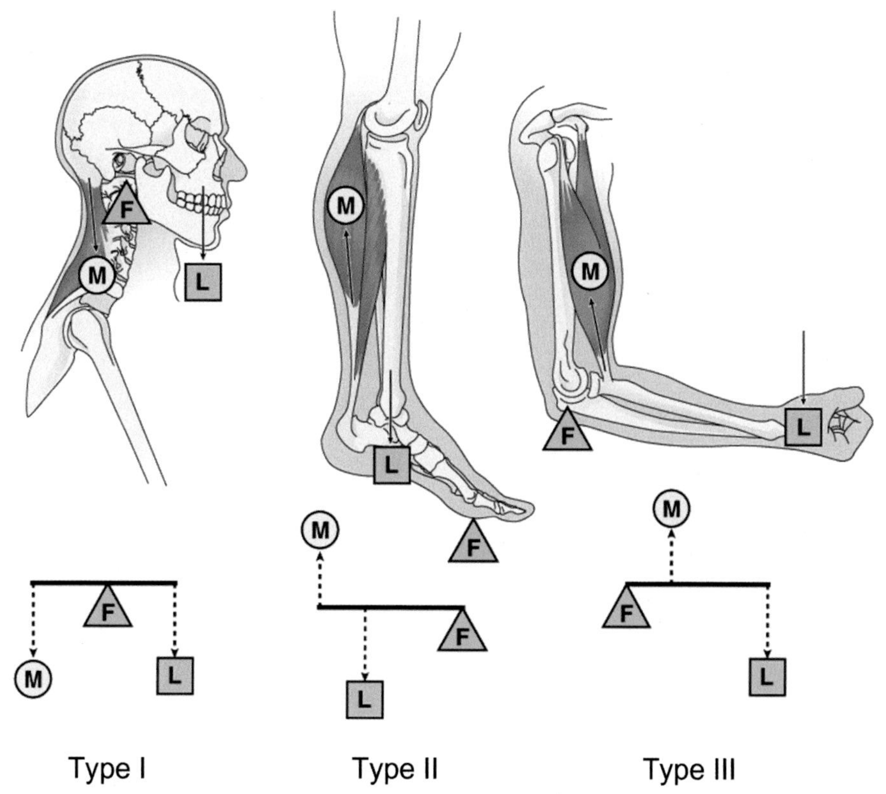

Figure 1.23 Muscles functioning as levers. *F*, fulcrum; *L*, load; *M*, muscle. *(From Carlson (2019), with permission.)*

rare in the human body, but is seen in the muscles at the back of the neck, as they counteract the weight of the head across the atlanto-occipital joint, where the neck joins the skull.

An often neglected aspect of whole-body muscle function is the role of strong sheets of connective tissue (**fascia**) in the transmission of muscular forces from one part of the body to another. Lateral connections between muscles through fascial sheets play a role in the smooth coordination of gross muscular movements. In addition, connections between the epimysium of muscles, their tendons and ligaments of joints often link distant parts of the body to one another (Fig. 1.24). Such connections can not only explain many body movements or postures, but they often account for a wide variety of chronic pains associated with the musculoskeletal system.

Almost any movement of a body part involves highly coordinated actions of several groups of muscles. Any flexion movement, for example, must be accompanied by the coordinated relaxation of the extensor musculature or that movement will not be effective. Such coordination is controlled through interactions of both the sensory and

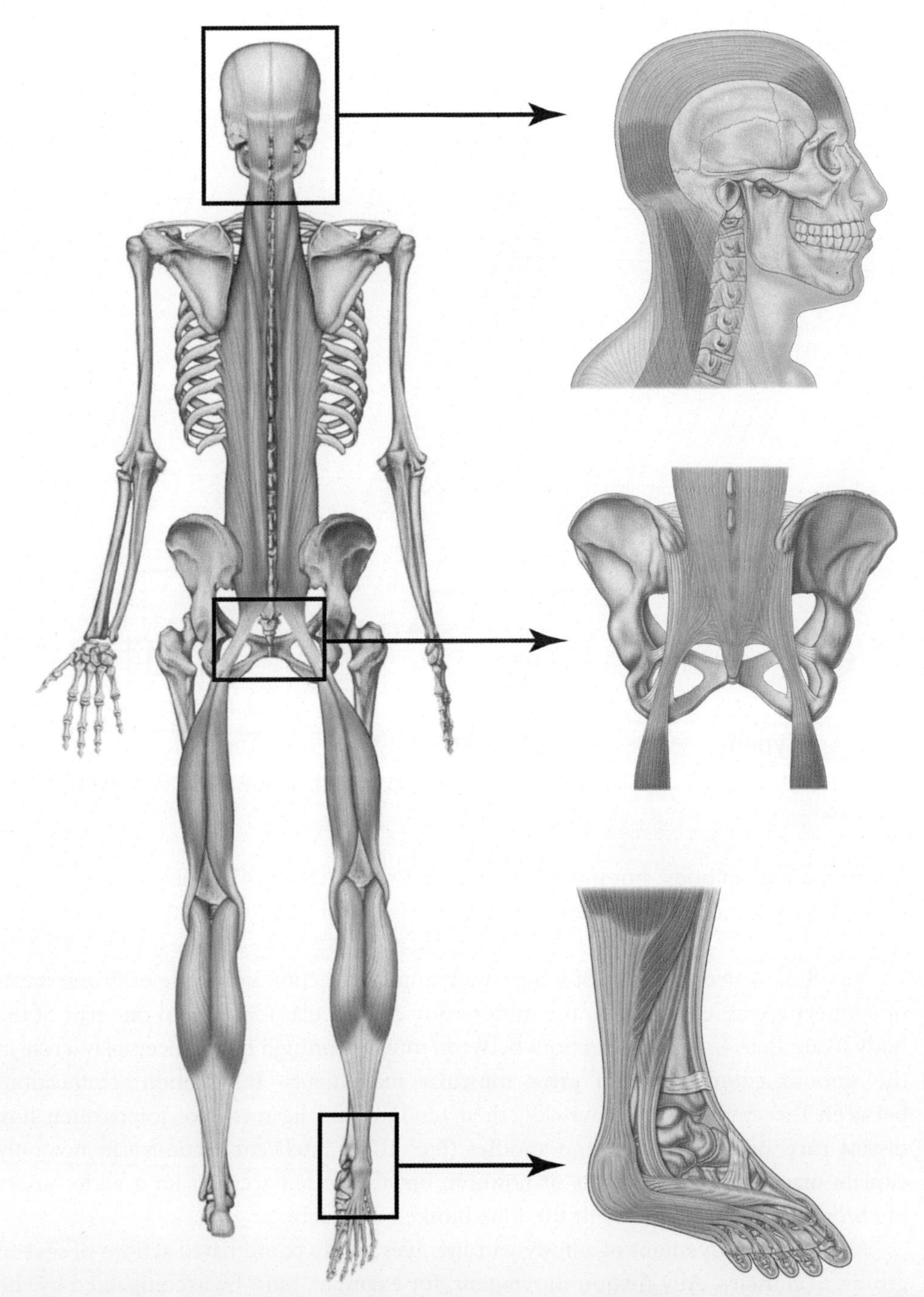

Figure 1.24 The anatomy trains concept, showing structural and functional connections (*in blue*) among a set of muscle groups throughout the body. *(From Myers (2014), with permission.)*

motor components of the nervous system in central processing areas, especially the cerebellum. Details of central coordination of body movements are beyond the scope of this book.

References

Boron WF, Boulpaep EL, editors. Medical physiology. 3rd ed. Philadelphia: Elsevier; 2017. p. 1247.

Carlson BM. The human body – linking structure and function. London: Academic Press; 2019. p. 417.

Charvet B, Ruggiero F, Le Guellec D. The development of the myotendinous junction. Muscles Ligs Tendons J 2012;2:53–63.

Gao QQ, McNally EM. The dystrophin complex: structure, function, and implications for therapy. Comp Physiol 2015;5:1223–39.

Gawor M, Proszynski TJ. The molecular cross talk of the dystrophin-glycoprotein complex. Ann N Y Acad Sci 2017;1412:62–72.

Glancy B, Hartnell LM, Malide D, Yu Z-X, Combs CA, Connelly PS, Subramaniam S, Balaban RS. Mitochondrial reticulum for cellular energy distribution in muscle. Nature 2015;523:617–20.

Janssen I, Heymsfield SB, Wang Z, Ross R. Skeletal muscle mass and distribution in 468 men and women aged 18-88 yr. J Appl Physiol 2000;89:81–8.

Lee YI, Thompson WJ. The vertebrate neuromuscular junction. In: Hilo AJ, Olson EN, editors. Muscle – fundamental mechanisms and disease. Amsterdam: Academic Press/Elsevier; 2012. p. 775–87.

Myers TW. Anatomy trains. 3rd ed. Edinburgh: Churchill Livingstone Elsevier; 2014. p. 317.

Narici M, Maganaris C. Muscle architecture and adaptations to functional requirements. In: Bottinelli R, Reggiani C, editors. Skeletal muscle plasticity in health and disease. New York: Springer; 2006. p. 265–88.

Peter AK, Cheng H, Ross RS, Knowlton KU, Chen J. The costamere bridges sarcomeres to the sarcolemma in striated muscle. Prog Pediatr Cardiol 2011;31:83–8.

Scharner J, Zammit PS. The muscle satellite cell at 50: the formative years. Skel Muscle 2011;1:1–13.

Schiaffino S, Reggiani C. Fiber types in mammalian skeletal muscle. Physiol Rev 2011;91:1447–531.

Yin H, Price F, Rudnicki MA. Satellite cells and the muscle niche. Physiol Rev 2013;93:23–67.

CHAPTER 2

Embryonic origins of skeletal muscle

Like many other parts of the body, skeletal muscle has a complex prenatal history that involves many developmental processes, such as proliferation, migration, differentiation, morphogenesis, and functional maturation. Many aspects of how a genetic blueprint is translated into the complex array of muscles in the adult body remain obscure and little investigated, whereas certain others, such as the molecular and cellular basis for muscle fiber differentiation, are now understood with a high degree of granularity. This chapter begins with the cellular origins of muscle and continues through the formation of anatomically identifiable muscles. The story of prenatal muscle development continues with a fetal growth phase, which continues into the postnatal period. In humans, childbirth itself is not a major milestone for the muscular system. This aspect of muscle development is covered in Chapter 3.

Origins and cellular migrations

If one wished to trace the ultimate cellular origins of muscle in the embryo, one could go as far back as the fertilized egg. At a practical level, however, a good place to begin is the somite. **Somites** are bricklike structures that are arranged in pairs on either side of the future central nervous system (Fig. 2.1). With the first somite pair appearing in the 25-day embryo, a new pair of somites is added every 4—5 h until by day 32 in humans, 30—31 pairs of somites are strung out alongside the developing spinal cord.

Somites are derived from the mesodermal germ layer of the embryo (Box 2.1) and represent the starting point for both the vertebral column and most of the trunk muscles. They take shape from strips of unsegmented **paraxial mesoderm** (see Fig. 2.4) that lie on either side of the neural tube. Seen in cross section, early somites are epithelial structures (Fig. 2.5), but a series of inductive signal-calling from the notochord and spinal cord causes the somite break up and to become subdivided into a **sclerotome** (future vertebra) and a **dermomyotome** (future dermis and muscle) (Fig. 2.6). The inductive signals are growth factors (Box 2.2), which exert profound effects on the somatic cells to which they are exposed (Fig. 2.5A and B). At the dorsal part of the somite, various members of the **Wnt** (Wnt—the mammalian equivalent of the gene wingless in *Drosophila*) family, arising from the dorsal part of the neural tube and the overlying ectoderm, stimulate the conversion of generic somatic cells into the dermomyotome. Ventrally, another powerful growth factor, **shh** (sonic hedgehog), induces ventral somatic cells to become sclerotome.

Muscle Biology
ISBN 978-0-12-820278-4, https://doi.org/10.1016/B978-0-12-820278-4.00009-8

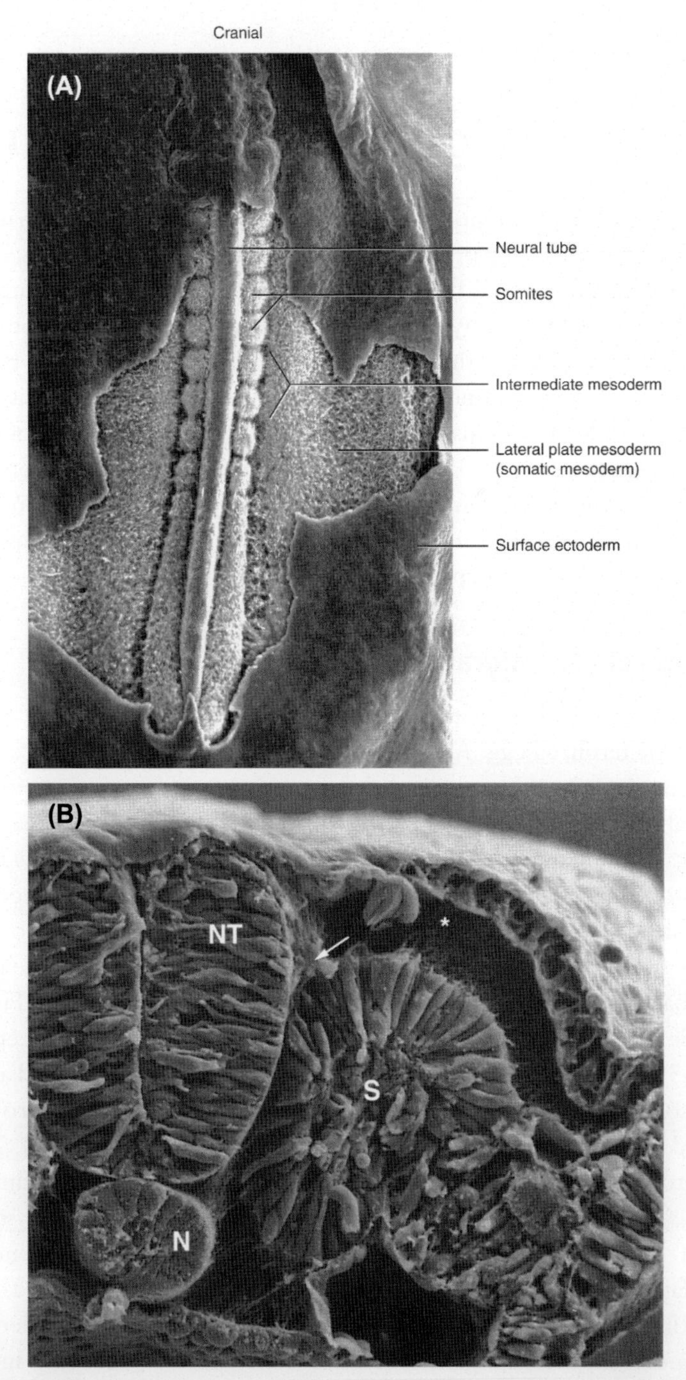

Figure 2.1 (A) Scanning electron micrograph of a chick embryo, showing rows of somites alongside the neural tube. (B) Scanning electron micrograph of a cross-sectioned chick embryo, showing somite (S) alongside the neural tube (NT). *N*, notochord. *((A) From Schoenwolf et al. (2015), with permission. (B) From Carlson (2019a), with permission.)*

BOX 2.1 An overview of early embryonic development

This chapter on muscle development begins with somites. Where do they come from? To answer this, an abbreviated version of early embryonic development is in order. The fertilized egg (zygote) is a single cell that contains essentially all the information necessary to form the body. About 24 h after fertilization, the zygote divides into two cells (**blastomeres**). Over the next 4 days (the period of cleavage), blastomeres continue to divide until by 5 days the embryo consists of a hollow sphere of cells with a smaller mass of cells (**inner cell mass**) on its inner surface (Figs. 2.2 and 2.3). The outer sphere of cells will form the placenta and other membranes that surround the developing fetus, whereas the inner cell mass will form the body of the embryo proper. Over the next week, the inner cell mass flattens and, through a process known as gastrulation, becomes transformed into three flat layers of cells. The upper layer is called **ectoderm**; the middle layer, **mesoderm**; and the lower

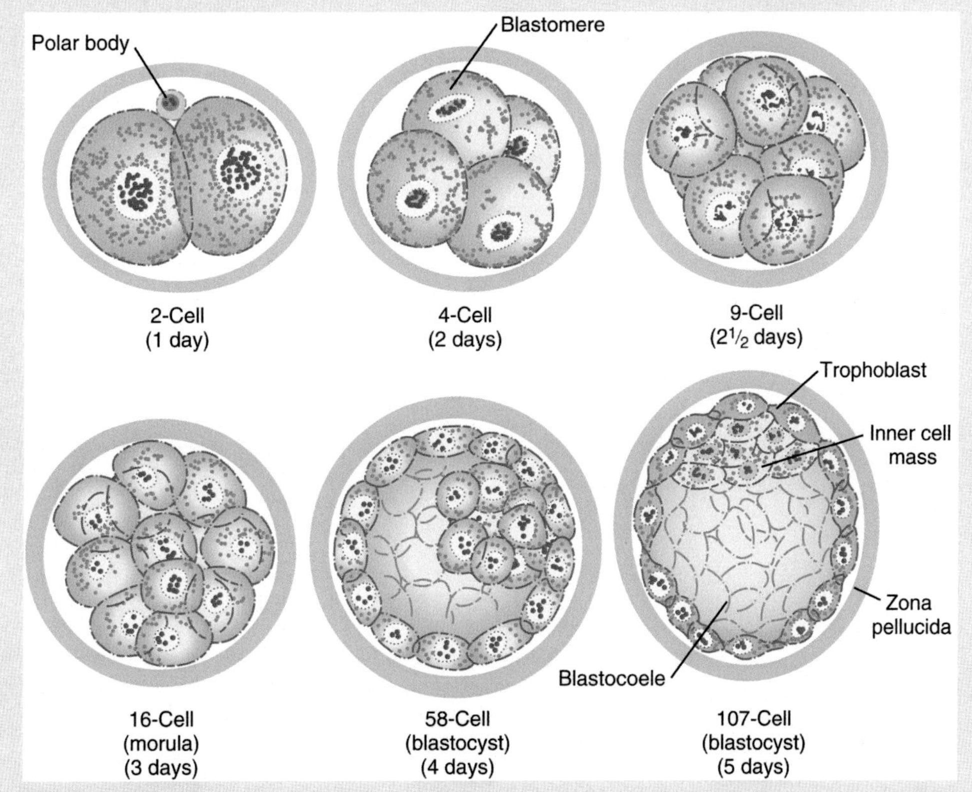

Figure 2.2 Drawings of human embryos during early stages of cleavage. *(From Carlson (2019a), with permission.)*

(Continued)

BOX 2.1 An overview of early embryonic development—cont'd

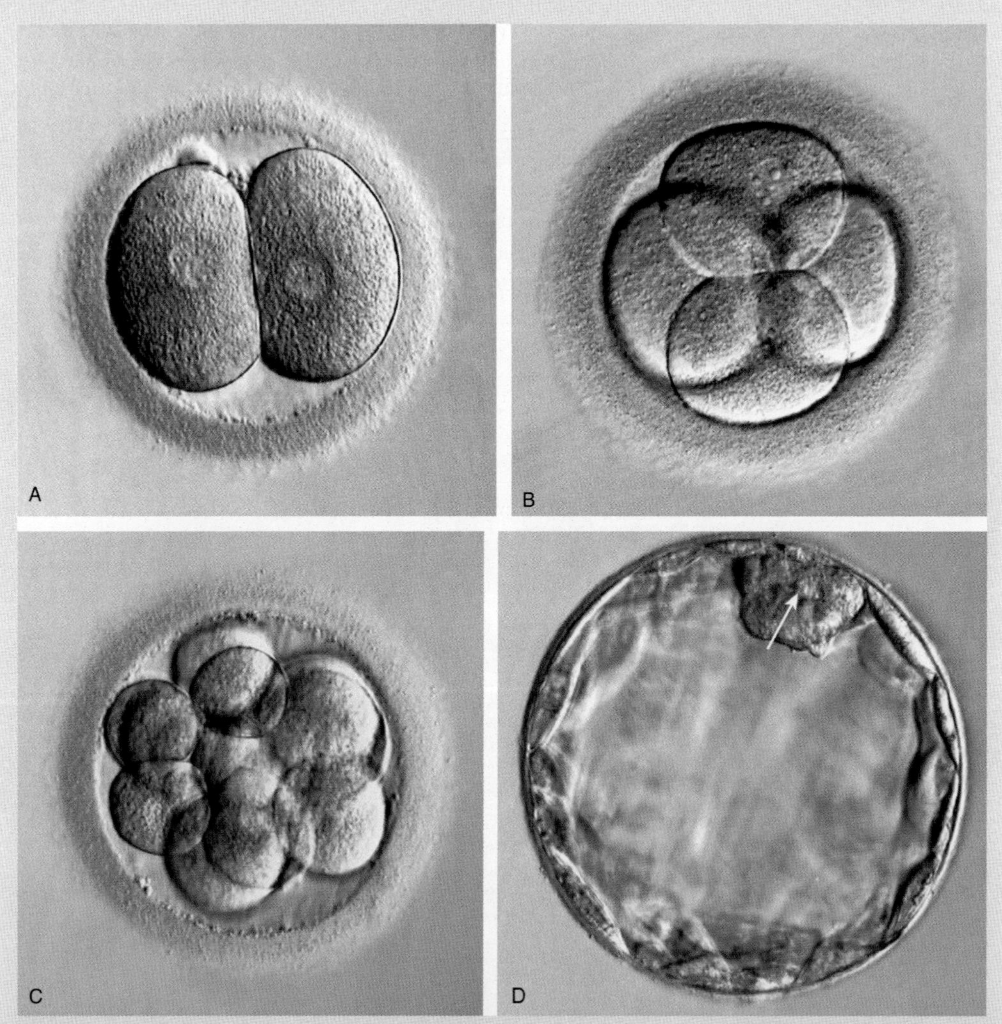

Figure 2.3 Photographs of early human embryos. (A) Two blastomeres. (B) Four blastomeres. (C) Twelve blastomeres. (D) Blastocyst with well-defined inner cell mass *(arrow)*, the cells that give rise to the body of the embryo. *(From Veeck and Janinovic (2003), with permission.)*

layer, **endoderm** (Fig. 2.4). Over the next week, the embryo will fold over to form a three-layered cylinder with ectoderm on the outside, mesoderm in the middle and endoderm forming a tube (the gut) inside. Through a signaling process, called induction, the dorsal ectoderm begins to form the precursors of the central nervous system (brain and spinal cord). Beneath and slightly off to the side, the underlying paraxial mesoderm, acting on other embryonic signals, forms pairs of blocklike somites, which account for the formation of almost all skeletal muscle in the body.

BOX 2.1 An overview of early embryonic development—cont'd

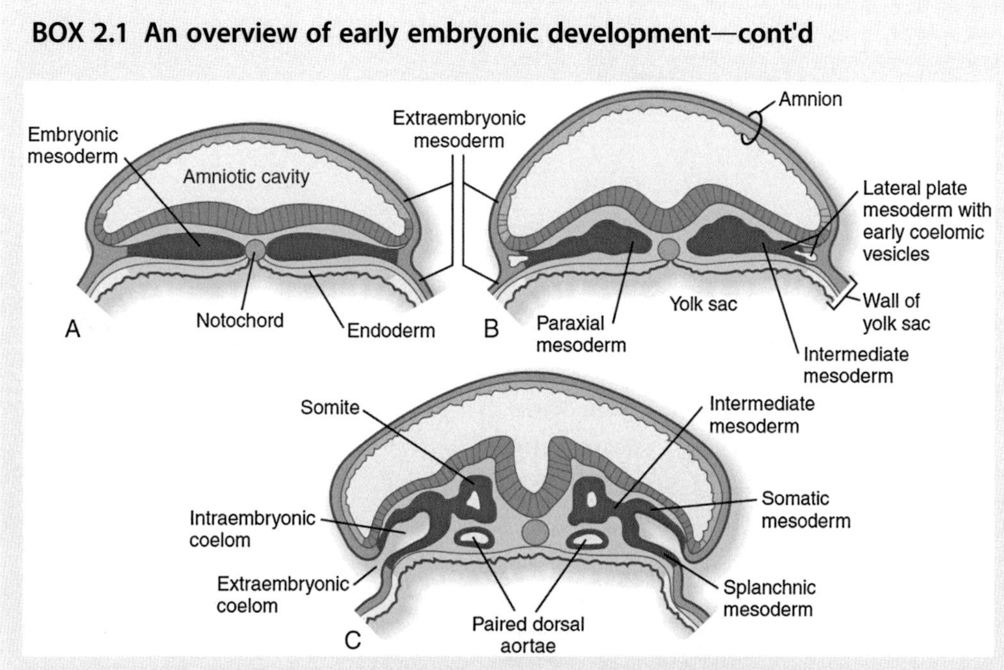

Figure 2.4 Cross sections through early human embryos, showing development of the ectodermal, mesodermal, and endodermal germ layers. *(From Carlson (2019a), with permission.)*

In response to further molecular signals, a wave of cells leaves the dermomyotome and initiates the formation of a separate layer, the **myotome** (see Fig. 2.6), which serves as the precursor of much of the musculature of the trunk. Cells of the dermomyotome express the transcription factor **Pax-3**, which is associated with a degree of flexibility concerning the future fate of these cells. As these cells become committed to **myogenesis** (muscle fiber formation), they stop expressing Pax-3 and instead produce **myogenic regulatory factors** (see below) that direct certain of these cells along a muscle-forming pathway. Cells in the middle third of the dermomyotome, which will contribute to the satellite cell population of muscle fibers, also express Pax-7—the major marker of quiescent satellite cells in postnatal life

Even as a myotome is beginning to form at the levels of the future limbs, groups of 100–300 **muscle progenitor cells** emigrate from the ventral border of the

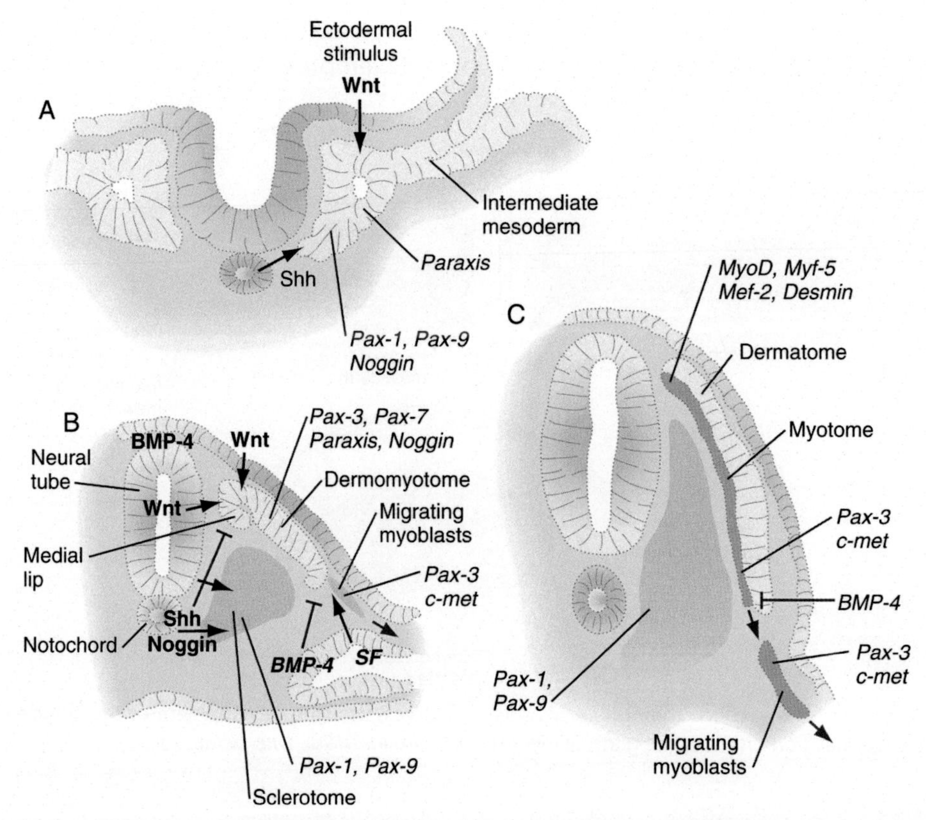

Figure 2.5 Molecular influences on the development of a somite. Signaling molecules are represented by *black arrows*. Inhibitory signals are represented by *red lines*. Genes expressed in responding tissues are indicated in italics. *BMP*, bone morphogenetic protein; *SF*, scatter factor; *Shh*, sonic hedgehog. *(From Carlson (2019a), with permission.)*

dermomyotome and move toward the site of limb formation (the limb bud—Fig. 2.8).[1] These cells are kept in a developmentally flexible state through the actions of another growth factor, bone morphogenetic protein (**BMP**), which is produced in the lateral plate mesoderm (see Fig. 2.5B). The migrating cells continue to express Pax-3 and do so until they arrive at the site of the future limb. Their migration is made possible through the attraction of a growth factor, **scatter factor** (also known as hepatic growth factor), which binds to a receptor, **c-met**, present on the surfaces of the migrating cells. Once

[1] That cells migrating out from somites form limb muscles were shown by experiments in which somites at the levels of the forming limbs were removed from chick embryos. In these cases, completely muscleless limbs with normal skeletons formed. Subsequent experiments involving specific cellular markers showed directly the pathway of migration of premuscle cells from the somite to the limb bud.

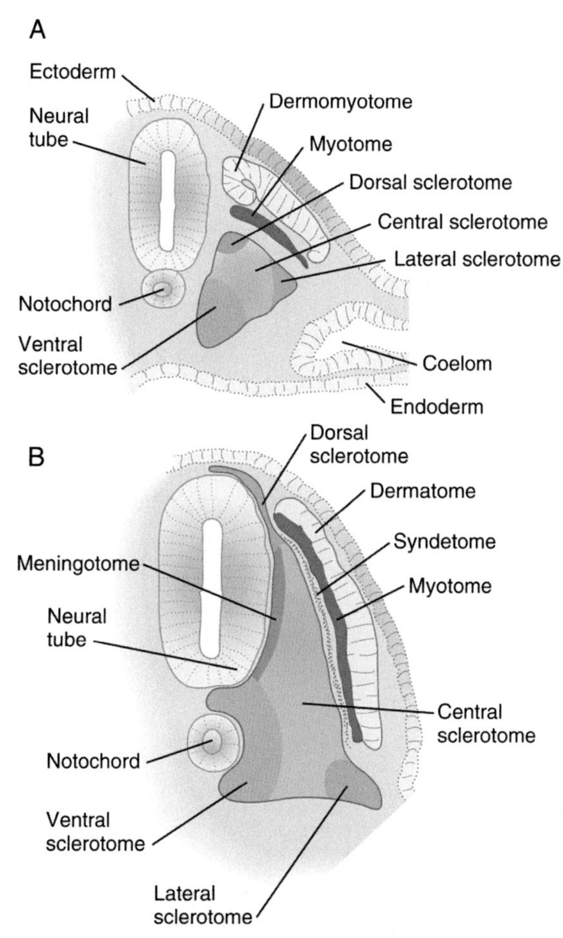

Figure 2.6 Organization of somites at earlier (A) and later (B) stages of development. *(From Carlson (2019a), with permission.)*

BOX 2.2 Growth factors and cytokines

Many of the developmental events involved in the embryology, growth, adaptation, and regeneration of muscle are initiated or guided by **growth factors** (signaling molecules). Growth factors are proteins produced by a large variety of cell types. They are typically secreted into the extracellular space (Fig. 2.7) and bind as **ligands** to **receptors** on nearby cells. Ligand-binding initiates an intracellular response through a **signal transduction pathway** that acts within the nucleus to influence the expression pattern of a gene or set of genes through their action on **transcription factors**.

(Continued)

BOX 2.2 Growth factors and cytokines—cont'd

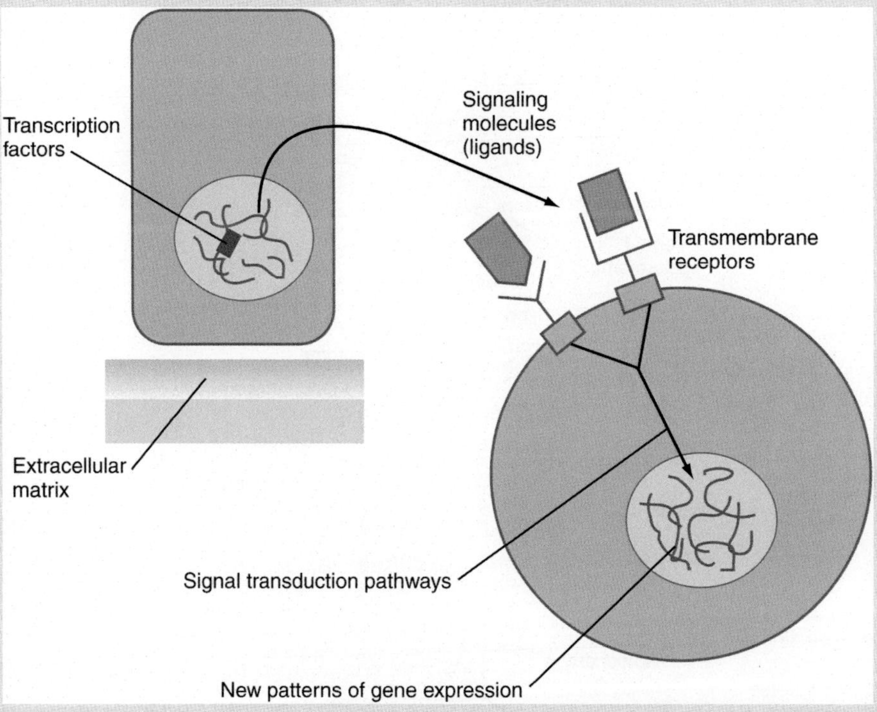

Figure 2.7 Schematic representation of types of developmentally important molecules and their sites of action. *(From Carlson (2019a), with permission.)*

Most growth factors are members of large families of related molecules, but even within a given family, individual growth factors may have widely diverse actions. In fact, a specific growth factor itself may have different functions at different times or in different locations. The major families of growth factors and members of these families mentioned in this book are listed in Table 2.1.

The **transforming growth factor-β** (TGF-β) superfamily contains a wide variety of quite different growth factors that affect almost every aspect of embryonic development. With respect to muscle, TGF-β, of which there are several isoforms, acts in concert with several other growth factors and cytokines to promote early myogenesis. After muscle damage, TGF-β plays an important role in the phase of the inflammatory response that leads to the activation of satellite cells. One of the most ubiquitous members of the TGF-β superfamily is the group of over 15 BMPs. BMP, originally discovered as an inducer of new bone after a fracture, plays many important roles in embryonic development, often acting as an inhibitor of other growth factors. One of its roles in embryonic muscle is to keep migrating premuscle cells developmentally flexible, so that they do not commit to differentiating before they arrive at their ultimate destination.

BOX 2.2 Growth factors and cytokines—cont'd

Table 2.1 Major families of growth factors.

Representative members	Function
Transforming growth factor-β (TGF-β) family	
TGF-β	Promoting early myogenesis
Bone morphogenetic protein (BMP)	Inhibitory effects on differentiation
Myostatin	Inhibitor of muscle growth
Fibroblast growth factor family	
Fibroblast growth factor (FGF)	Often stimulates cell division
Wnt family	
Wnt	Induction of muscle
Hedgehog family	
Sonic hedgehog (shh)	Involved in induction of ventral somite

The FGF family is another large one, with over 20 members so far identified. It was originally discovered as a molecule stimulating the division of fibroblasts in vitro. FGFs are important embryonic inducers of structures as diverse as limbs, hair, teeth, parts of the brain, and many glands throughout the body. In many developing systems, various FGFs function to keep cells in the cell cycle, rather than differentiating. This is one of their principal functions in satellite cell biology.

The **Wnt** family, consisting of almost 20 members, is heavily involved in a wide variety of inductive processes. It plays a critical role in inducing cells of the early somite to enter a myogenic pathway, and in postnatal life, Wnts are key players in the early phases of regenerative myogenesis.

The **hedgehog** family is a small one, with only three members, but one of them, **sonic hedgehog** (shh) is one of the most powerful inductors in the embryonic body. Among its many functions, shh directs critical aspects of the formation of major parts of the brain and spinal cord, limbs, teeth, hair, the lungs, and the retina. Shh, in conjunction with Wnts, works to subdivide the early somites into skeletal and muscular components.

Cytokines is an umbrella term given to a very large number of signaling molecules produced by macrophages and cells of the immune system. Overall, they are involved in coordinating the progression of the inflammatory response to injury and, in the case of muscle, in making the transition from inflammation to early phases of regenerative myogenesis, especially the activation of satellite cells.

Figure 2.8 Migration of muscle precursor cells (*arrows*) from somites.

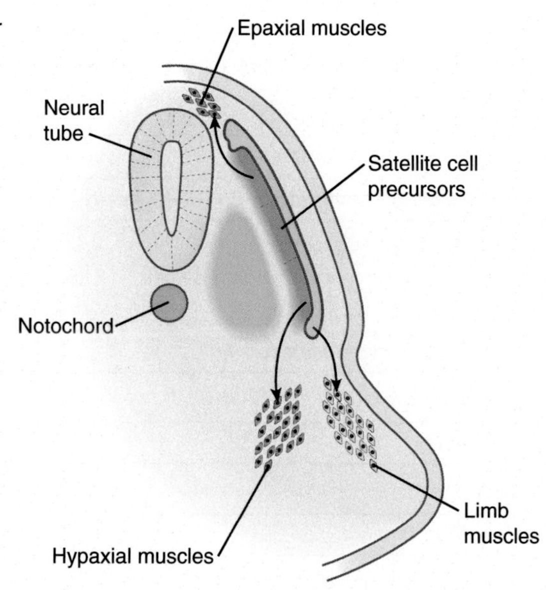

these migrating cells have arrived in the limb bud, they become molecularly committed to becoming muscle-forming cells by expressing specific myogenic regulatory factors. Some members of this stream of muscle progenitor cells reverse direction once they have arrived in the limb bud and migrate back into the future trunk region, where they will ultimately produce the muscles of the limb girdles and the perineum. At other appropriate levels along the upper body axis, streams of cells arising from the ventrolateral edge of the dermomyotome migrate toward their final location as intrinsic muscles of the tongue and diaphragm.

Meanwhile, the myogenic cells that remain within the myotome of the somite proceed to form two important groups of trunk muscles. Those in the dorsal third of the myotome begin to form the group of **epaxial muscles**—the deep muscles of the back that lie alongside the vertebral column (see Fig. 2.8). Those cells in the ventral third of the myotome project ventrally to form the **hypaxial muscles**—those of the abdominal wall and the intercostal muscles. Two other superficial muscles of the trunk, the trapezius and sternocleidomastoid muscles, have an unusual origin that only makes sense from their evolutionary history.[2] They do not arise from somatic precursor cells, but rather from cells originating in the lateral plate mesoderm (see Fig. 2.4B). Their myogenic

[2] These muscles are evolutionary descendants of the cucullaris muscle, a muscle found in submammalian vertebrates that connects the dorsal skull to the dorsal pectoral muscles.

transcription factors belong in the same family as those that guide the development of heart muscle instead of those that guide myogenesis in the rest of the trunk and limbs.

The central third of the dermomyotome does not give rise to major muscle groups. Instead it produces the progenitors of satellite cells—the stem cells that are vital for the growth and regeneration of skeletal muscle fibers. These satellite cell precursors supply both axial and limb muscles. Another subset of the satellite cell population also arises from paraxial mesoderm, but from cells resident in the early dorsal aorta. One of the surprising results of research on muscle development has been that muscles of the head and neck develop under quite a different set of rules from those that guide the development of trunk and limb muscles (Box 2.3).

BOX 2.3 Early development of muscles of the head and neck

A big surprise to muscle biologists has been the recognition of how many features of craniofacial muscles differ greatly from those of trunk and limb muscles. Many of these differences stem from their evolutionary origins. Most muscles in the face (muscles of facial expression and those that open and close the jaws) are both phylogenetically and ontogenetically connected with the **branchial (gill) arch** systems that are of critical importance to aquatic vertebrates. This humble evolutionary origin generated a group of cells in the early embryo of higher vertebrates known as **cardiopharyngeal mesoderm**. Initially, the molecular signatures of these cells are the same, but in response to signals from different environmental regions, some of the precursor cells within the cardiopharyngeal mesoderm develop into heart muscle and others become precursors for the pharyngeal musculature.

Those muscles involved in opening and closing the jaw are associated with the first branchial arch, which during evolution became transformed into the jaw apparatus. The superficial muscles of the face are derived from precursors connected to the second branchial arch, and even in humans, they migrate to the face from the neck region during embryological development. Cellular precursors for these muscles arise for the most part from unsegmented paraxial mesoderm in the cranial region. These precursor cells move into the branchial arches, and from there they spread out over the forming face and neck. The intrinsic muscles of the tongue are derived from somites. Their precursor cells migrate from the most cranial somites into the developing tongue. The sternocleidomastoid and trapezius muscles are evolutionarily derived from the cucullaris muscle (see p. 44), and their precursor cells originate in lateral plate mesoderm. The extraocular muscles are unusual in many respects. Their origin still remains debated, but they follow their own molecular pathway during myogenesis (see Fig. 2.9). Even in adulthood, they contain some unique myosins (see Table 1.2) and have both structural and functional properties that are different from those of most other muscles. It has been known for many years that the connective tissue that invests muscles of the head and neck originates from neural crest[2a] ectoderm rather than the mesoderm that is associated with skeletal muscles throughout the rest of the body.

(Continued)

BOX 2.3 Early development of muscles of the head and neck—cont'd

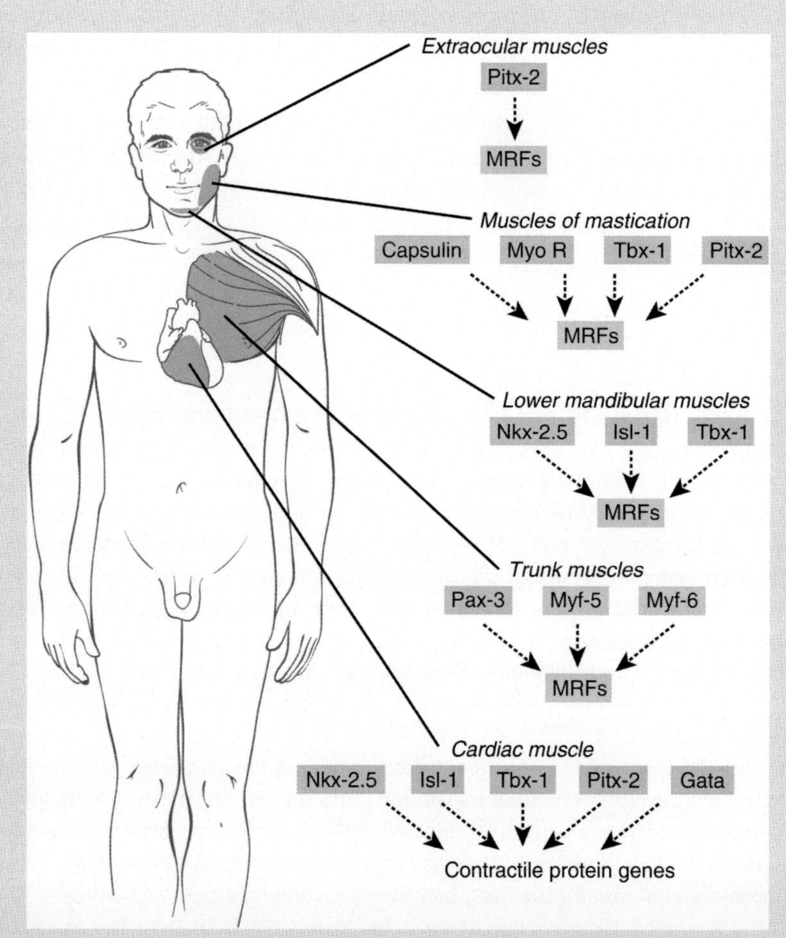

Figure 2.9 Molecular controls in the determination and differentiation of the various groups of striated muscle in the body, based on studies in the mouse. *(From Carlson (2019a), with permission.)*

The latest surprise has been the recognition that there are significant differences between the myogenic regulatory factors that guide the differentiation of muscle fibers of craniofacial muscles versus those of the trunk and limbs (Fig. 2.9). Whereas almost all trunk and limb muscles employ roughly the same repertoire of transcription factors and myogenic regulatory factors, the differentiation of several groups of craniofacial muscles is controlled by several different sets of regulatory molecules. With this new understanding, it is now becoming possible to understand basis underlying the different anatomical distributions of several muscle diseases.

[2a] Neural crest is an aggregation of mesenchymal cells that emigrate from the closing neural tube in the early embryo. Arising from the ectodermal layer, they spread out as three streams along the sides of the forming head and become closely associated with the precursor cells of the craniofacial muscles as they move toward their final locations in the head.

Morphogenesis of a muscle

Once they have arrived in the limb bud, the muscle progenitor cells spread throughout the proximal regions of the limb bud. At that point, their progression along the muscle-forming lineage involves ceasing the production of Pax-3 and expressing myogenic regulatory factors (see below). As the mesenchyme in the center of the limb bud undergoes condensation to form the primordia of the skeletal elements, the muscle progenitor cells aggregate into two separate groups to form the precursors (**blastemas** or **common muscle masses**) of the flexor and extensor muscle groups. Local environmental influences play a significant role in the positioning and organization of the cells within the common muscle masses. A growth factor, **Wnt-6**, produced by the ectoderm covering the limb bud, promotes myogenesis while at the same time inhibiting the formation of cartilage in the areas where muscle will form. This influence results in the placement of the skeleton in the central core of the limb, with the muscles forming alongside the skeleton. The distal end of the limb bud, which is an active zone of cell proliferation and outgrowth, does not contain muscle progenitor cells, possibly due to an inhibitory influence of the growth factor **BMP**, which is present in high concentrations in the distal limb bud. Signals from the overlying ectoderm of the limb bud (Wnt-6) are required for the differentiation of the muscle progenitor cells into actual muscle fibers. In the absence of the ectoderm, cartilage and connective tissue continue to form, but muscle fails to appear.

One of the major mysteries in muscle development is how individual limb muscles take form from the common muscle masses. The common muscle masses begin to split from proximal to distal regions in well-defined patterns that presage the formation of individual muscles. It is known that neither muscle progenitor cells nor their descendants possess the intrinsic information that determines their morphogenetic fate. This has been shown through experiments in which somites from different regions of the body have been transplanted in place of somites that would ordinarily supply the limbs with muscle cells. The myogenic progenitor cells from the transplanted somites, which would have formed completely different muscles in their normal locations, instead took part in the formation of normal limb muscles. In essence, even though they have the capacity to differentiate into perfectly normal muscle fibers, muscle cells themselves are morphogenetically ignorant and follow the morphogenetic lead of their surrounding connective tissue.

It is now known that morphogenetic control of limb muscle development resides in the connective tissue cells that are resident in the limb bud. These cells express the transcription factor **Tcf-4**, but how (or if) this is translated into shaping the common muscle masses into individual muscles remains poorly understood. The initial formation of individual muscles appears to be independent of physical connections with tendons or other components of the limb bud. There is evidence suggesting that local blood vessels secrete a growth factor (**platelet-derived growth factor**) that stimulates the growth of connective tissue in the areas that would separate one muscle from another, thus providing a basis for their separation. The earliest stages of not only muscle, but also skeletal and tendon development, occur autonomously, and only later are connections made.

Tendons arise from local Tcf-4$^+$ limb bud mesenchyme, and their initial appearance is independent of muscles. In fact, when muscleless limbs were created in chick embryos (see p. 40, footnote [1]), these limbs were found to have essentially normal early tendons. In most cases, tendons begin to form slightly after the primordia of both skeletal elements and muscles begin to take shape. A typical limb tendon is initially induced by a growth factor (**fibroblast growth factor** [**FGF**]) produced by the overlying ectoderm, but in the hand or foot, cartilage is needed for tendon induction. After their initial induction, pretendon cells then express the transcription factor **scleraxis** (**Scx**), which guides and is necessary for their early development and differentiation. These tendon primordia then secondarily attach both to the ends of the muscle primordia and the appropriate skeletal elements (see p. 28).

The connection between a mature tendon and bone is called an **enthesis**, and it consists of a succession of tissue types from tendon to fibrocartilage to mineralized cartilage and finally to bone (see Fig. 1.20). Collagen fibers extending from the tendon become embedded within the bony substance, which provides the strength that prevents the rupture of the tendon—bone connection when a muscle contracts maximally. The initial tendon-bone connection consists of a sharp transition between tendon tissue and cartilage, the precursor of a long bone. Only after muscle loading begins to occur in the fetal period and even after birth does the formation of a true enthesis begin to take place. Rather than just growing out of the forming bone, the cells making up the enthesis arise from a separate pool of cells that require the presence of muscle in order to begin the process of enthesis formation. These cells insert themselves into the tendon—bone junction and respond to the mechanical forces of the developing muscle by producing **fibrocartilage**, a tissue that contains elements of both tendon and cartilage.

Myogenesis

While muscles as anatomical entities are taking shape, profound changes are also occurring at the level of muscle cells. At a structural level, the most significant change is the formation of elongated multinucleated muscle fibers from a population of mononuclear cells. As this process is taking place, profound intracellular changes—both molecular and structural—equip the developing muscle fiber with the means to contract. In a broader sense, myogenesis does not only take place at the single cell level. While all of the internal processes of muscle fiber differentiation are occurring, the developing muscle cells are also beginning to organize at the level of muscle tissue. This involves both interactions among the muscle fibers themselves and their association with other types of cells, in particular, the cells of nerves, blood vessels, and the connective tissue that surrounds a muscle. At an even higher level, the sculpting of the entire muscle as an individual muscle with its unique internal architecture also takes place.

Development of a single muscle fiber. Once myogenic progenitor cells have settled down in a limb bud, they follow a complex pathway of differentiation on their way to becoming mature multinucleated muscle fibers. From a purely morphological perspective, muscle precursor cells, which are morphologically indistinguishable from other mesenchymal cells, begin to assume a spindle shape, at which point they are called **myoblasts** and slightly later, **myocytes** (Fig. 2.10). Then individual myocytes begin to line up in a side-by-side fashion and fuse, forming multinucleated **myotubes** with chains of nuclei lined up in a central core. As the myotubes mature, they begin to produce well-ordered bundles of contractile proteins surrounding the nuclear chains. With further maturation, the centrally located nuclei of the myotube begin to migrate to the periphery, at which point it can be called a **muscle fiber**. Embryonic muscle fibers elongate through the fusion of mononuclear myoblasts to either end.

As it is maturing, the myotube and early muscle fiber produce a thin noncellular basement membrane that encases the entire muscle fiber. Between the muscle fiber and its surrounding basement membrane lies a population of mononucleated stem cells (**satellite cells**; see Fig. 1.2), which represent the basis for growth of the muscle fiber during normal development and the regeneration of new muscle fiber material after injury.

The development of a muscle fiber from undifferentiated muscle progenitor cells takes place under tight molecular control. The first set of controls, four **myogenic regulatory factors** (Myf-5, MyoD, myogenin, and Myf-6), is involved in making the commitment to becoming a muscle fiber and guiding its further development. Myogenic regulatory factors are **transcription factors** that activate the cascade of molecular events that commits a cell to becoming a muscle cell. Subsequent controls guide the synthesis of the contractile proteins and their regulatory elements.

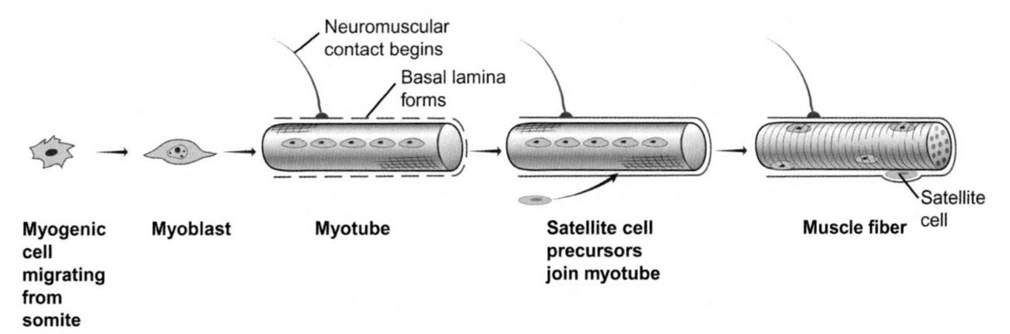

Figure 2.10 Embryonic development of a skeletal muscle fiber from a myogenic cell. Many myoblasts fuse to form a single myotube, which then matures to form a muscle fiber. *(From Carlson (2007), with permission.)*

While muscle progenitor cells are migrating to the limb bud from the somites under the influence of Pax-3 and c-Met, other molecular influences keep them from expressing myogenic regulatory factors and becoming prematurely committed to forming muscle. Once they have settled down in the limb bud, the muscle progenitor cells downregulate the molecules (Pax-3 and c-Met) that guided their migration and begin to activate the molecules of the myogenic cascade. This begins with the expression of **Myf-5**, which acts to commit these cells to a myogenic fate (Fig. 2.11). This, and a later contribution by **MyoD**, activates **myogenin**, which is heavily involved in the transition of myoblasts to multinucleated myotubes. A final myogenic regulatory factor, **Myf-6**, is active during later stages of differentiation of the muscle fiber.

Another important regulatory layer involved in myogenesis is the large group of **microRNAs** (miRNAs). miRNAs are small (20–25 nucleotides) RNAs that are heavily involved in mRNA silencing and in the posttranscriptional regulation of gene expression. To date, several thousand different miRNAs have been identified, and among these more than 150 are known to be expressed in muscle. In skeletal muscle, miRNAs have been found to modulate almost every stage of myogenesis and adaptation. Some miRNAs, for example, miR-31 and miR-29c play a role in promoting quiescence in satellite cells, whereas miR-133 promotes proliferation. Still others support differentiation, switching of fiber types from slow to fast and regeneration of muscle fibers.

As myoblasts begin to fuse to form multinucleated myotubes, much of the molecular activity turns to the synthesis of contractile proteins (actin and myosin), their regulatory elements (troponin and tropomyosin), and the other structural proteins that keep them in

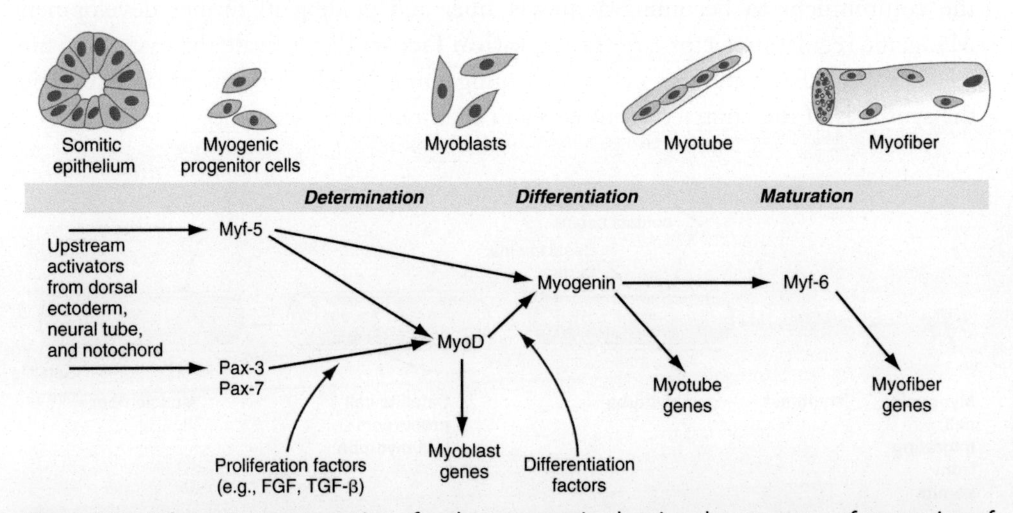

Figure 2.11 Schematic representation of early myogenesis, showing the sequence of expression of myogenic regulatory factors and other influences on the myogenic process. *FGF*, Fibroblast growth factor; *TGF-β*, transforming growth factor-β. *(From Carlson (2019a), with permission.)*

place. These are discussed in Chapter 3. In the embryo and during the growth period, the contractile proteins undergo **isoform transitions**. **Isoforms** are different forms of the same protein, which are adapted for particular stages in one's life cycle. A good non-muscle example is hemoglobin, where embryonic and fetal isoforms have different oxygen-binding capacities from adult hemoglobin. These are adaptations for the conditions of intrauterine life. Isoforms of contractile proteins are characterized by different contractile properties of developing muscle fibers.

Development (histogenesis) of muscle tissue. A tissue is an aggregation of different cell types, all of which work to accomplish some main function, which in the case of a muscle is to produce a contraction. The process of muscle histogenesis involves both the muscle cells themselves and several types of associated cells.

The first muscle fibers to form in a developing limb are called **primary muscle fibers**, and the process is called **primary myogenesis**. The muscle fibers formed during primary myogenesis are responsible for establishing the fundamental pattern of the muscles, and they initially take shape independently of any neural influence. Primary myogenesis, however, is not sufficient to support the growth of a muscle. Much of muscle growth is accomplished by a later wave of **secondary myogenesis** (Fig. 2.12), a process by which a different set of myoblasts, operating under a different set of molecular controls from those that guide the formation of primary muscle fibers, forms new muscle fibers alongside the primary muscle fibers (Fig. 2.13). Although the formation of secondary muscle fibers also involves the fusion of myoblasts, this process is in general nerve-dependent. In the absence of motor nerve fibers, most secondary myogenesis does not

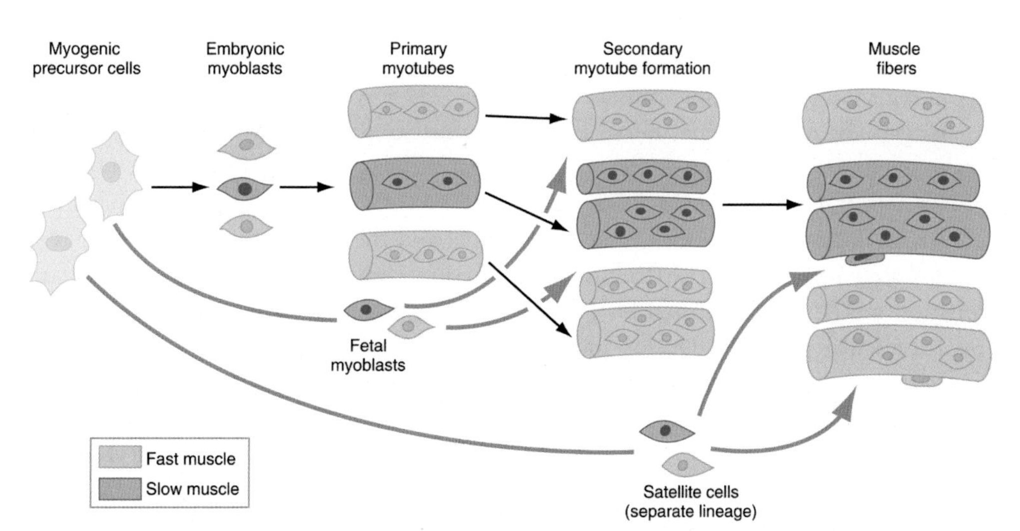

Figure 2.12 Stages in the formation of primary and secondary muscle fibers. A family of embryonic myoblasts contributes to the formation of primary myotubes, and fetal myoblasts contribute to secondary myotubes. *(From Carlson (2019a), with permission.)*

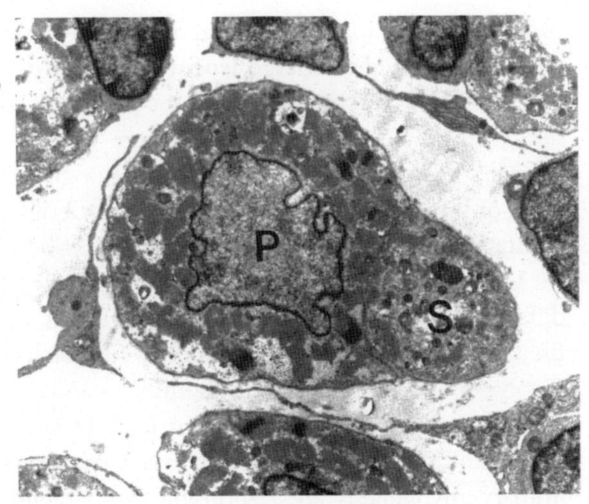

Figure 2.13 Electron micrograph of a secondary muscle fiber (S) forming alongside a primary muscle fiber (P). *(Courtesy of D Evans.)*

occur, whereas in primary myogenesis, muscle fibers do form initially without innervation, although innervation is required for their further development.

All muscle fibers are not created equal, and even in the embryo fast and slow muscle fibers are present. Initially, all primary muscle fibers are slow in character, but over time some of them begin to form fast embryonic myosins, especially at the periphery of the muscle. For many years, it was thought that innervation by fast or slow motor nerve fibers determined the character of the embryonic muscle fibers, but more recent research has shown that the muscle fiber type is determined even before a muscle fiber is contacted by nerves. It now appears that fast or slow motor nerve fibers are attracted to the corresponding type of embryonic muscle fiber.

Regardless of their type, nerve fibers and muscle fibers follow a similar pattern of making connections during embryogenesis. A mature muscle fiber is connected to a single branch of a nerve fiber at an area called the **neuromuscular junction**, which involves specializations of both the nerve terminal and the part of the muscle fiber lying immediately beneath it. (The structure and functional aspects of a mature neuromuscular junction are described on p. 144. Familiarity with this will aid in understanding how neuromuscular junctions develop in the embryo.)

In the embryo, muscle fibers prepare for their eventual innervation by prepatterning a specific site in their middle region that is attractive to the tips of growing motor nerve fibers. Although initially spread throughout a developing muscle fiber, **acetylcholine receptors** (AChR) begin to cluster roughly midway between the two ends of the muscle fiber. The clustering of AChR, structures fixed in the membrane of the muscle fiber, is facilitated and stabilized by another protein, **rapsyn**, which in turn interacts with two other membrane proteins, **Lrp4** (lipoprotein receptor-related protein) and **MuSK**

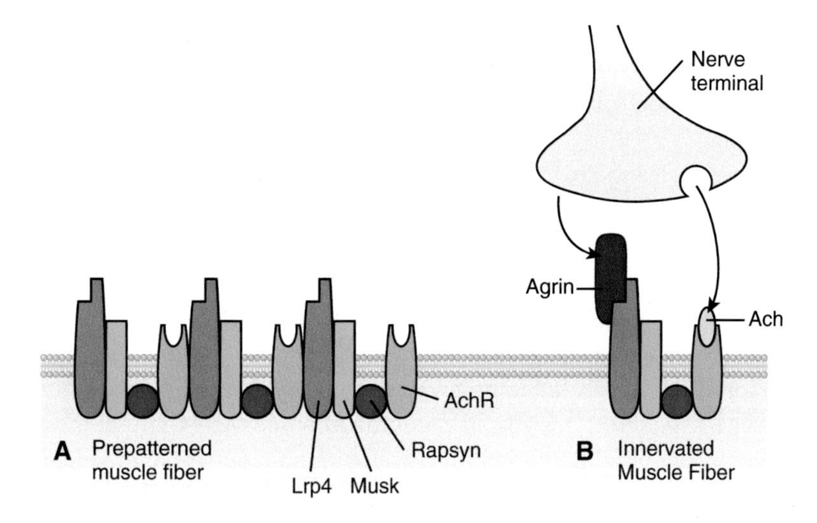

Figure 2.14 Molecular aspects of neuromuscular junction formation. The postsynaptic apparatus (A) Forms first and is then stabilized by agrin, produced by the nerve terminal (B). When the neuromuscular junction becomes functional, acetylcholine (Ach), released from synaptic vesicles in the nerve terminal, binds to AchR.

(muscle-specific kinase), both of which are heavily involved in stabilizing the neuromuscular junction (Fig. 2.14). At this point, before the nerve arrives, their main function is assisting the clustering of AChR.

As ingrowing nerves approach the muscle fibers, they are attracted to the accumulated AChR. The nerve fiber terminals produce **agrin**, a protein that inserts itself into the basal lamina that is interposed between the muscle fiber and the nerve terminal (see Fig. 2.14). Agrin binds to Lrp4, which interacts with MuSK to further stabilize the neuromuscular junction. Acting in reverse, Lrp4 plays an important role in maturation of the nerve terminal.

One unusual feature of early muscle development is that initially a muscle fiber is often innervated by as many as six motor nerve terminals. Over the course of several weeks, a pruning process takes place, with the result that ultimately a single muscle fiber is innervated by only a single nerve fiber. How this occurs and why muscle fibers are initially multiply innervated is subject to considerable speculation, but definitive answers have remained elusive.

As muscle fibers form and surround themselves with a basement membrane of their own making, they become closely associated with connective tissue cells that envelop them. This connective tissue, consisting of **collagen fibers** and the **fibroblasts** that make them, forms a wispy endomysium that surrounds individual muscle fibers. At a grosser level of partitioning, sheets of **perimysium** enclose and separate bundles of muscle fibers (see Fig. 1.1). Surrounding the entire muscle is a capsule of connective tissue, called the **epimysium**. These connective tissue components play a vital role in transmitting the forces of contraction by individual muscle fibers to the tendons and the entire limb.

A growing muscle requires a blood supply, and from the early stages in the formation of a muscle, developing muscle fibers become associated with networks of capillaries that supply them with nutrients and remove metabolic wastes. As the muscle tissue develops, differences in the blood supply to different types of muscle fibers become apparent. Slow muscle fibers, which when mature rely greatly upon oxidative metabolic processes for their function, become invested with much more robust capillary networks than are fast muscle fibers.

Muscle spindles follow a different developmental course. They first appear during the early fetal stage (e.g., 11−12 weeks in the human) while the extrafusal muscle fibers are still in the early stages of differentiation. Developing spindles are highly dependent upon innervation. Intrafusal fibers are first contracted by sensory nerve fibers and require them for their further differentiation. Nuclear bag fibers are the first to form. Only later, after nuclei have accumulated in the bag, do motor nerves make connections with the polar ends of the intrafusal fibers (where the contractile material is located). Motor nerves first contact nuclear bag fibers; only later do motor nerves connect with nuclear chain fibers. In the absence of sensory innervation, spindles do not form. As the intrafusal muscle fibers differentiate, a thin capsule of connective tissue begins to form around the spindle.

Development of entire muscles. Entire muscles begin to take shape shortly after muscle progenitor cells have settled into the flexor and extensor common muscle masses and splitting of the muscle masses has begun to occur. Most limb muscles form along a proximodistal gradient, with the most mature part of the muscle at the proximal end. This gradient follows a general developmental gradient in the limb bud that applies to skeletal elements, as well.

Most limb muscles are recognizable as discrete entities as soon as the splitting of the common muscle masses has occurred. At this point, the muscles have not yet made connections with the developing tendons. Very early in their developmental history, the essential elements of the internal architecture of most muscles are already set in place. Particularly obvious is the orientation of the primary muscle fibers, which anticipates that of the fully mature muscle. Nowhere is this more obvious than the intrinsic muscles of the embryonic tongue. The adult tongue contains bundles of muscle fibers that are oriented along each of the three Cartesian axes. Such orientation is already clearly evident as muscle is first forming in the tongue.

In the developing limb, cartilaginous models of the skeletal elements begin to differentiate first. This is followed by the formation of muscles. Tendons and muscles initially form from independent cellular sources. Tendon precursor cells, called **tenoblasts**, arise form lateral plate mesoderm, whereas myoblasts migrate into the limb bud from somites. These two types of cells begin to associate very early when tenoblasts begin to secrete a fine fibrillar matrix in the vicinity of the myoblasts at the end of the premuscle mass. As the myoblasts fuse to become myotubes, a basal lamina begins to form around the nascent myotubes. The terminal basal lamina becomes loosely interdigitated with the extracellular matrix secreted by the tenoblasts. As the myotubes develop the ability to contract, the tension from these contractions aligns the collagen fibers of the tenocyte—myotube interface into a parallel configuration. With increasing strength of muscle contractions, the muscle fiber—tendon interface develops the fingerlike configurations that characterize the mature myotendinous junction (see p. 28).

With the passage from the embryonic to the fetal stage, the muscles begin to contract. This mechanical activity plays a role in the final shaping of both the muscle itself and the tendons that attach it to bone, as well as the tendon—bone interface itself. The fetal period is characterized by rapid growth and functional maturation of the muscles.

References

Altana V, Geretto M, Pulliero A. MicroRNAs and physical activity. MicroRNA 2015;4:74—85.

Buckingham M, Rigby PWJ. Gene regulatory networks and transcriptional mechanisms that control myogenesis. Dev Cell 2014;28:225—38.

Burden SJ, Yymoto N, Zhang W. The role of MuSK in synapse formation and neuromuscular disease. Cold Spring Harbor Perspect Biol 2013;5:a009167.

Burden SJ, Huijbers MG, Remedio L. Fundamental molecules and mechanisms for forming and maintaining neuromuscular synapses. Int J Mol Sci 2018;19:490—509.

Carlson BM. Human embryology and developmental biology. 6th ed. St. Louis: Elsevier; 2019. p. 480.

Chai J, Pourquié O. Making muscle: skeletal myogenesis *in vivo* and *in vitro*. Development 2017;144: 2104—22.

Charvet B, Ruggiero F, Le Guellec D. The development of the myotendinous junction. a review. Muscles Ligs Tends J 2012;2:53—63.

Deries M, Thorsteinsdóttir S. Axial and limb muscle development: dialogue with the neighborhood. Cell Mol Life Sci 2016;73:4415—31.

Huang AH. Coordinated development of the limb musculoskeletal system: tendon and muscle patterning and integration with the skeleton. Dev Biol 2017;429:420—8.

Michailovici I, Eigler T, Tzahor E. Crainiofacial muscle development. Curr Top Dev Biol 2015;115:3—29.

Murphy M, Kardon G. Origin of vertebrate limb muscle: the role of progenitor and myoblast populations. Curr Top Dev Biol 2011;96:1—32.

Musumeci G, Castrogiovanni P, Coleman R, Szychlinska MA, Salvatorelli L, Prenti R, Magro F, Imbesi R. Somitogenesis: from somite to skeletal muscle. Acta Histochem 2015;117:313—28.

Pu Q, Patel K, Huang R. The lateral plate mesoderm: a novel source of skeletal muscle. In: Brand-Saberi B, editor. Vertebrate myogenesis. Berlin: Springer-Verlag; 2015. p. 143—63.

Pu Q, Huang R, Brand-Saberi B. Development of the shoulder girdle musculature. Dev Dynam 2016;245: 342—50.

Sambasivan R, Kuratani S, Tajbakhsh S. An eye on the head: the development and evolution of craniofacial muscles. Development 2011;138:2401—15.

Scaal M, Marcelle C. Chick muscle development. Internet. J Dev Biol 2018;62:127—36.

Subramanian A, Schilling TF. Tendon development and musculoskeletal assembly: emerging roles for the extracellular matrix. Development 2015;142:4191—204.

Tzahor E. Head muscle development. In: Brand-Saberi B, editor. Vertebrate myogenesis. Berlin: Springer-Verlag; 2015. p. 123—42.

Wigmore PM, Evans DJR. Molecular and cellular mechanisms involved in the generation of fiber diversity during myogenesis. Int Rev Cytol 2002;216:175—232.

CHAPTER 3

Normal muscle growth

Normal muscle growth begins in the fetus, with the formation of secondary myotubes. It continues uninterruptedly, although with different aspects of growth being emphasized, into the early postnatal phase and through the postpubertal period until overall growth of the body ceases in the late teens or early twenties in the case of humans. An increase in muscle size can and does occur after that time, but it is normally an adaptation to exercise, a topic which will be covered in Chapter 4.

As seen through a broad lens, the dominant theme of human prenatal muscle growth is **hyperplasia**—an increase in the number of muscle fibers. Most studies suggest that by the time of birth, a body contains almost all of the muscle fibers that it will ever possess, although it does not necessarily imply that some degree of turnover of muscle fibers may not occur. After birth, muscle growth focuses on the **hypertrophy** (increase in size) of existing muscle fibers.

In many respects, the growth of a muscle fiber can be reduced to two fundamental processes: (1) the progressive accumulation of myonuclei through the incorporation of satellite cell progeny and (2) the balance between the synthesis and degradation of muscle fiber proteins. Both of these involve a delicate dance that involves a variety of intrinsic and environmental factors that influence the progression and intensity of these events.

Prenatal muscle fiber formation

The formation of primary muscle fibers from somite-derived cells in the embryo was covered in Chapter 2 (p. 35), and the formation of secondary muscle fibers was also briefly introduced (see Figs. 2.12 and 2.13). Primary muscle fibers form in greatest numbers toward the end of the first quarter of pregnancy, whereas most secondary muscle fibers form during the second and third quarters (Fig. 3.1). Secondary muscle fibers first form alongside primary muscle fibers as very thin nerve-dependent fibers. From their initial formation, secondary muscle fibers must increase in both length and diameter in order to keep up with overall growth of the fetus, which increases in length over 10-fold between the time when secondary myogenesis begins and birth. Growth in length allows a greater **excursion** (change in length) of a muscle fiber, whereas an increase in diameter increases the contractile force that the muscle fiber can exert.

By the time secondary myogenesis is underway, the belly of an embryonic muscle has become attached to its tendons of origin and insertion, and the tendons have become connected to bone (see p. 28). As the bones of the limbs (and other skeletal elements) grow in length, they exert mechanical tension upon the muscles to which they are

Muscle Biology
ISBN 978-0-12-820278-4, https://doi.org/10.1016/B978-0-12-820278-4.00003-7

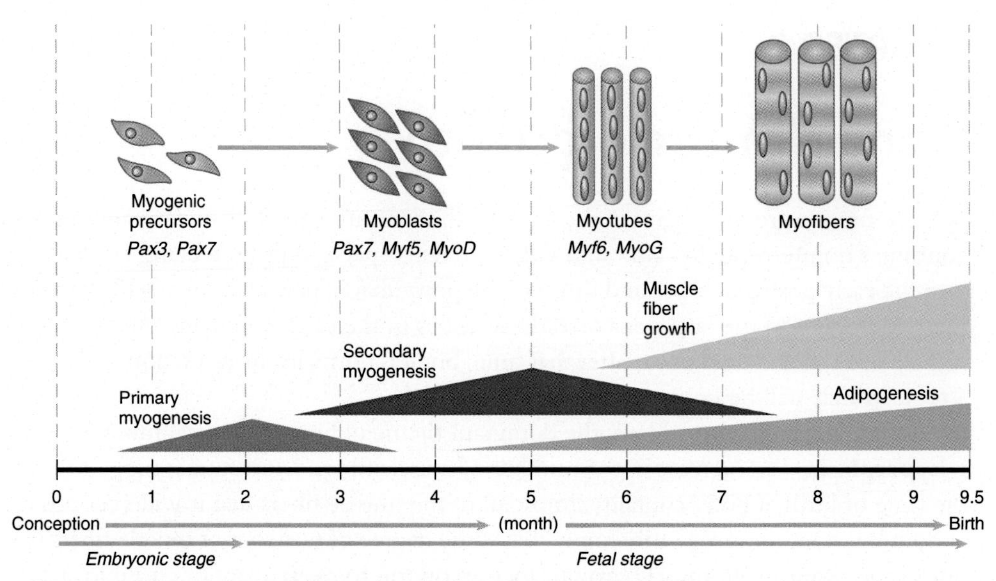

Figure 3.1 Phases of prenatal muscle growth. Active transcription factors are indicated in italics.

attached. This mechanical tension is then translated into growth in length of the secondary muscle fibers. Growth in length can be accomplished in two ways. In the fetus, muscle fibers elongate principally by adding myoblastic cells to their growing ends. The myoblasts fuse to the terminal portions of the secondary muscle fibers. The other mechanism of tension-induced growth is the addition of sarcomeres, mainly at the ends of an existing muscle fiber. This can occur both before and after birth. Growth in diameter of a muscle fiber is accomplished mainly by the production of new contractile proteins. These self-assemble into new sarcomeric components in parallel with those already present in the muscle fiber.

Development of muscle fiber types

Fetal, as well as adult, muscle fibers are not all uniform. Rather, even at this stage, they can be divided into fast and slow types based upon the types and isoforms of myosin proteins that they are producing (see p. 60). Different contents of myosin isoforms constitute a significant basis for the classification of muscle fibers into different types. After a review of the structure of a generic myosin molecule, this section will follow the development of the various major types of muscle fibers.

The myosin molecule. A generic mammalian muscle myosin molecule consists of two main components—heavy chains and light chains (Fig. 3.2). The heavy chain has two protruding heads connected to long α-helically coiled tail segments. The heads are the components that interact with actin during muscle contraction through their

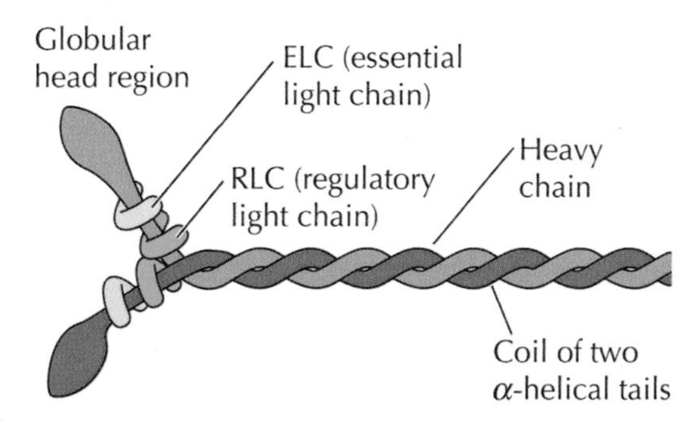

Figure 3.2 Structure of a thick filament, showing the myosin heavy and light chains.

ATPase activity, and the tails form the bulk of the thick filaments (see p. 14). There are two types of **light chains**, each of which enwraps the neck region that connects the globular heads to the tails. One of the light chains is designated **LC1** (essential, or alkali) and the other, **LC2** (regulatory). Details of their function are still being worked out, but they essentially modulate the functions of heavy myosin during cross-bridge formation. There are isoform variants of both myosin heavy chains (MHC) and light chains. These appear both at different stages of development of a single muscle fiber and among different types of muscle fibers at any given developmental stage.

Developmental isoforms of myosin. Skeletal muscle fibers possess different isoforms of MHC at different stages of their development. In addition, the various types of muscle fibers also possess different MHC isoforms. To complicate matters even further, there are different patterns of progression of MHC isoforms between primary and secondary muscle fibers. During transitional periods, individual muscle fibers commonly contain more than a single myosin isoform.

When primary muscle fibers first form, they all contain the embryonic form of MHC_{emb}. They soon also begin to produce MHC_{β}, a myosin isoform that characterizes Type I (slow) muscle fibers, even in the adult. By about the time of birth, MHC_{emb} is lost, but in Type I fibers derived from primary myogenesis, MHC_{β} persists as the characteristic MHC (Fig. 3.3). Those primary muscle fibers destined to become fast muscle fibers follow a somewhat different course. Like future slow fibers, these also express MHC_{emb} and MHCβ sequentially, but as the time of birth approaches, MHC_{emb} and MHCβ are replaced by a neonatal myosin, MHC_{neo}. After birth, these muscle fibers can differentiate into any of the standard adult muscle fiber types. The differentiation of primary muscle fibers is independent of any influence by motor nerve fibers.

Secondary muscle fibers initially all express MHC_{emb}, followed by MHC_{neo} (see Fig. 3.3). Around birth, MHC_{emb} has largely disappeared, whereas MHC_{neo} persists for a while after birth. After birth, MHCβ begins to accumulate in those secondary fibers

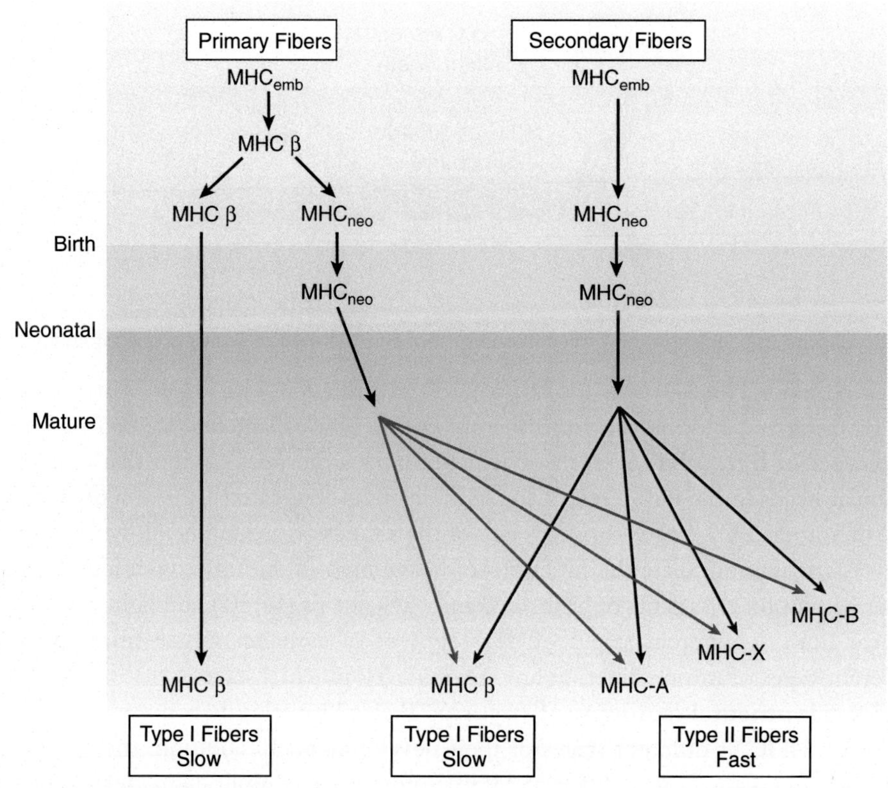

Figure 3.3 Progression of myosin isoforms in skeletal muscle development. *emb*, embryonic; *MHC*, myosin heavy chain; *neo*, neonatal. *(After Schiaffino et al. (1996).)*

destined to become slow fibers. The future Type II muscle fibers then replace the MHC_{neo} with MHC isoforms appropriate for types II A, X, or B muscle fibers. The myosin light chains also undergo isoform transitions over time, but these will not be covered here. Regardless of the types of myosins that they contain, fetal and neonatal muscles contract slowly. Only later, as they mature, do the contraction times of fast muscle fibers speed up.

Experiments have identified three major factors that underlie these developmental changes in MHC isoforms. One is the transition from polyneuronal innervation of muscle fibers to single innervation, which allows a more concentrated neural influence upon the character of the muscle fiber. A second is the imposition of mechanical load upon developing muscle. This is particularly important in the development of slow muscle fibers in postural muscles. A third is the influence of thyroid hormone, which plays a prominent role in the expression of fast myosin genes.

Even in postnatal life, muscle fiber type is not fixed. By natural or experimental means (e.g., through exercise or by changing their motor nerve supply), muscle fibers can be stimulated to change their functional type and the myosin isoforms that they contain. In essence, these environmental influences beyond the muscle fiber itself cause changes in gene expression that result in shifts of the myosin isoforms found in the muscle fiber.

Satellite cells and postnatal muscle growth

Satellite cells (see Fig. 1.2) were first discovered in 1961 and at that time could only be identified through electron microscopy. Satellite cells are small nondescript cells, located between a muscle fiber and its overlying basal lamina. In a typical nongrowing muscle fiber, there is roughly one satellite cell for every 25–50 myonuclei. For many years, their function remained enigmatic. Now it is known that they play vital roles in normal muscle growth, hypertrophy, and also regeneration. In effect, satellite cells represent a population of adult stem cells whose main function is to maintain a stable population of muscle fibers that is appropriate for maintaining the functions demanded of them. There appear to be two main subpopulations of satellite cells—one that is directed toward supporting the growth of muscle fibers and another that focuses on selfrenewal of the satellite cell population itself.

The importance of satellite cells during periods of rapid muscle growth can be inferred from their abundance. In both rats and pigs, the number of satellite cells in relation to the total number of nuclei (myonuclei + satellite cell nuclei) beneath the basal lamina surrounding a muscle fiber is roughly 35% at birth and then drops during the phase of rapid muscle growth to about 4%–5% in mature muscle. Then during the aging process, that percentage gradually diminishes to about 1%–2% in very old age. As the percentage of satellite cells falls during the early growth period, the number of myonuclei per muscle fiber correspondingly increases—about five times during the first 3 weeks of life in a rat.

Numbers of satellite cells also differ between fast (type II) and slow (Type I) muscle fibers. They are more abundant in slow than in fast muscle fibers. During the aging process, satellite cell abundance declines more rapidly in fast than in slow muscle fibers.

The quiescent satellite cell. In normal nongrowing muscle, satellite cells are actively maintained in a prolonged state of quiescence until they are called upon to support the further growth or regeneration of that muscle fiber. Quiescent satellite cells are characterized by their expression of the transcription factor, Pax-7. Continued expression of **Pax-7** and also the activity of Notch (see below) are required to maintain satellite cells in the quiescent state. In their quiescent (in contrast to dormant) state, satellite cells are molecularly programmed to be in a state of readiness to become activated when the need arises. Maintenance of this state is complex, with the expression of as many as 500 genes involved.

One of the main regulators of the quiescent state is the **Notch pathway**. This is a pathway that a number of developing systems use to prevent the premature or inappropriate differentiation of neighboring cells. In the case of muscle, the **Notch** receptor

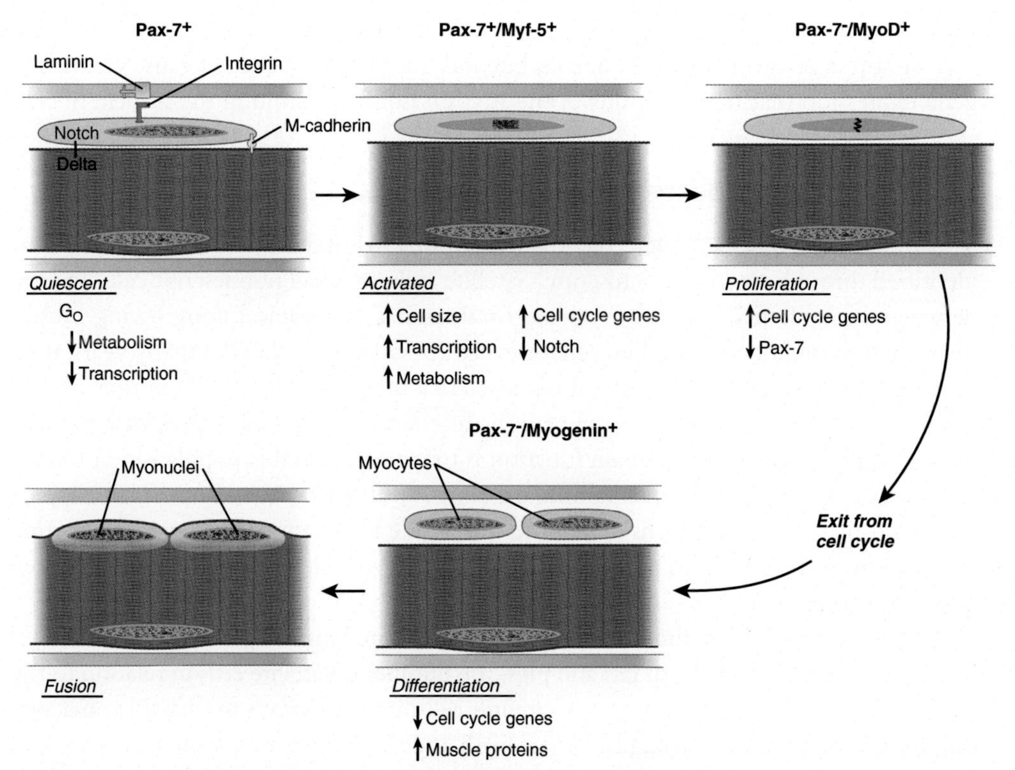

Figure 3.4 Stages in the development of a satellite cell from quiescence to incorporation into a muscle fiber.

molecule is expressed on the surface of satellite cells, and its ligand, **Delta**, is found on the surface of the underlying muscle fiber (Fig. 3.4). In their quiescent state, the overall metabolism of satellite cells is maintained at a low level, and genes that bring cells into the cell cycle are inactive.

Satellite cells occupy a unique niche between a muscle fiber and its basal lamina. There is increasing evidence that satellite cells communicate with both the muscle fiber and the basal lamina in different ways. The plasma membrane on the basal side (facing the basal lamina) of the satellite cell expresses an **integrin** (see p. 27) which connects it to the laminin component of the basal lamina. The plasma membrane on the apical surface (facing the muscle fiber) expresses the cellular adhesion molecule, **M-cadherin**, through which it connects to the membrane of the muscle fiber. On the basal lamina side, satellite cells can influence events relating to the connective tissue surrounding the muscle fibers. On the muscle fiber side, continuing interactions maintain the quiescent state until something disturbs the equilibrium. Then activation of satellite cells can occur.

As a muscle fiber undergoes early postnatal growth, satellite cells are called into play. More myonuclei are needed to fulfill the transcriptional demands of the growing muscle fiber. These demands include the need to produce more contractile proteins through RNAs produced by the myonuclei. According to one theory (**myonuclear domain theory**), a given myonucleus can only direct the production of a certain amount of cytoplasmic proteins. To satisfy the needs of the growing muscle fiber, more nuclei must be added to meet these demands. Research has shown that the number of myonuclei per muscle fiber increases postnatally. Because myonuclei cannot divide, incorporation of satellite cell nuclei into the growing muscle fiber is a way to meet this demand. A first step in this process is activation of the satellite cells. As a muscle fiber grows postnatally, the myonuclear domain increases, especially in Type IIB and Type IIX fibers. In contrast, the myonuclear domain in slow muscle fibers (Types I and IIA) remains more constant.

The activated satellite cell. Although it is well known that products of injury activate satellite cells, the environmental factors that activate satellite cells during normal growth remain much less well defined. Before activation, satellite cells have exited from the cell cycle (in the G_0 state), do not express myogenic regulatory factors, and have low transcriptional and metabolic activity. Important early stages in activation are removal of the Delta/Notch inhibition that characterizes quiescent satellite cells (see above) and activation of some of the genes that allow them to return to the cell cycle (see Fig. 3.4). Even though they still express Pax-7, activated satellite cells begin to express **Myf-5**, the first level myogenic regulatory factor (see Fig. 2.11). Overall, transcriptional and metabolic activity also increase.

The major result of satellite cell activation is a return to the cell cycle and the division of individual satellite cells. At this point, they no longer express Pax-7, but now express the myogenic regulatory factor, **MyoD**. Proliferation of satellite cells increases their number, but what the daughter cells resulting from these mitotic divisions will become is critical. During normal muscle growth, there are two main options. The daughter cells could continue their myogenic trajectory and fuse with the muscle fiber or they could remain as committed satellite cells. If they all followed the myogenic pathway, the original population of satellite cells would ultimately become exhausted. On the other hand, if they all remained in the satellite cell lineage, there could be no contribution to the growth of the muscle fiber.

A major determinant of the future fate of activated satellite cells is their orientation as they divide between the muscle fiber and the basal lamina. If the orientation of the mitotic spindle is parallel to these structures (**planar division**, Fig. 3.5), there is a high likelihood that both daughter cells resulting from that division will be of the same type, most likely remaining as satellite cells. On the other hand, if there is an **apicobasal division** (one daughter cell next to the muscle fiber and the other next to the basal lamina), there is a high likelihood that the two daughter cells will have different developmental fates. The one located next to the muscle fiber will turn into a myoblast and become incorporated into the muscle fiber, whereas the one closest to the basal lamina will retain its properties as a satellite stem cell. After proliferation, those daughter cells that

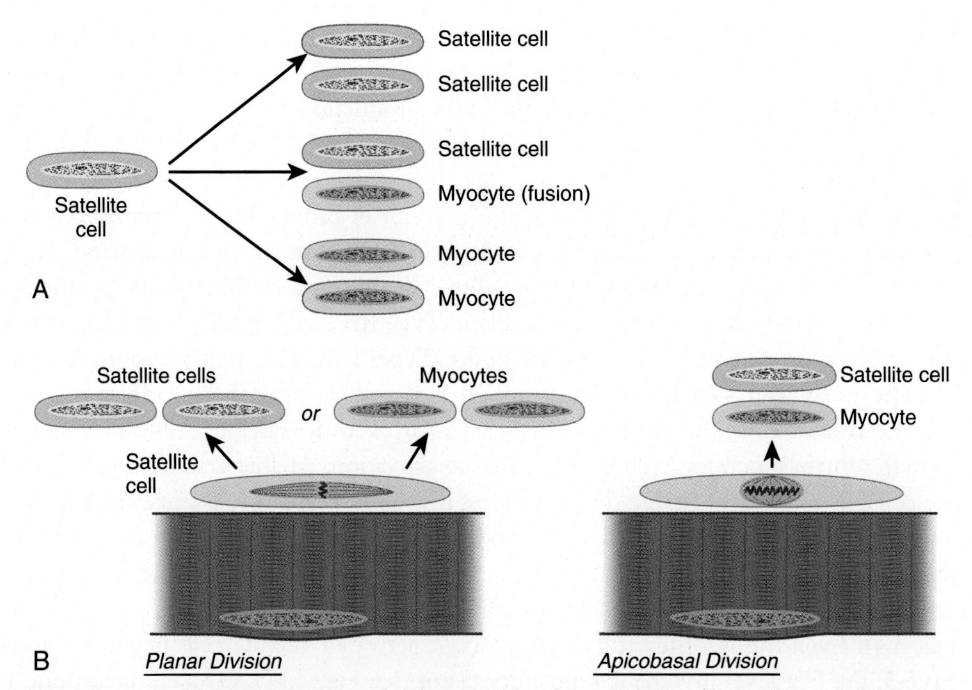

Figure 3.5 (A) Developmental options for the progeny of the division of a satellite cell. (B) Developmental consequences of planar versus apicobasal divisions of a satellite cell.

will remain in the stem cell pool reexpress Pax-7, but do not express any of the myogenic differentiation factors. They pull out of the cell cycle and return to the G_0 state and reestablish a Notch/Delta relationship with the muscle fiber as regular satellite cells. Satellite cells expressing Myf-5 are more likely to continue down the myogenic line, whereas those that do not continue to maintain stemlike properties. Those daughter cells that become myogenic will express **myogenin** but have lost the expression of Pax-7. They will then prepare to fuse with their associated muscle fiber.

Satellite cell fusion. After their commitment to becoming myogenic, satellite cell progeny in mammals become migratory. Most commonly, they migrate along their associated muscle fiber, but they can also penetrate the basal lamina and seek out other muscle cells. The latter behavior is more common after muscle injury. At some point, the migrating myoblastic cell settles down and prepares to fuse with a muscle fiber. Fusion of satellite cell progeny with a muscle fiber is a surprisingly complex process, with many molecular pathways involved. Much of what we know is derived from studies on *Drosophila* muscle, but the general sequence of fusion events seems also to occur in mammalian muscle.

Fusion begins with adhesion of what is called a **fusion competent myoblast** to the sarcolemma of a muscle fiber. A large number of adhesion-related molecules have been

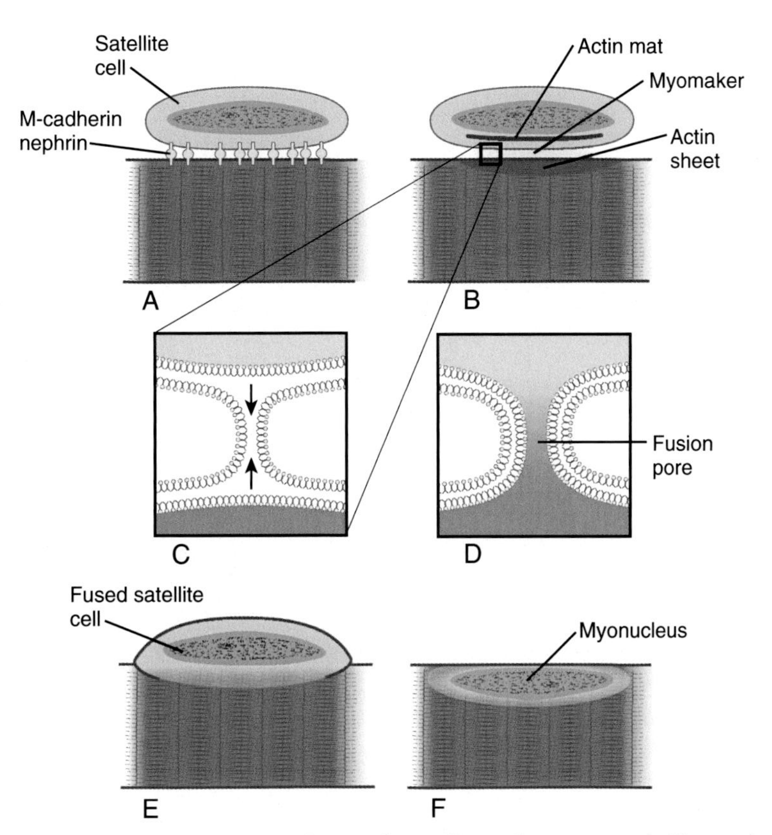

Figure 3.6 Successive stages (A–F) in the fusion of a satellite cell onto a muscle fiber and its ultimate fate as a myonucleus.

identified in mice. Two prominent ones are **nephrin** and M-cadherin, which is also involved in the adhesion of quiescent satellite cells to the underlying muscle fiber. After the initial adhesion event, actual fusion between the myoblast and the muscle fiber begins.

One of the earliest steps in the fusion process is the formation at the site of fusion of a sheet of actin filaments beneath the surface of the muscle fiber (Fig. 3.6). At the corresponding side of the myoblast, a dense mat of actin forms. Then the plasma membranes of the 2 cells make contact and membrane remodeling begins. The first stage of membrane remodeling is merging of the outer layer of the plasma membranes of the 2 cells. This is followed closely by a similar merging of the inner layer of the 2 cells, resulting in the creation of a **fusion pore** that connects the cytoplasm of the 2 cells. Fusion pores expand into areas of the muscle fiber where the actin sheet breaks down locally until ultimately the entire contents of the myoblast are incorporated into the existing muscle fiber. When fusion is complete, the nucleus of the incorporated satellite cell is now capable of directing the synthesis of additional proteins needed for growth of the muscle fiber.

Two recently discovered small proteins, **myomaker** and **myomerger**, are critical to the fusion process. Although their exact roles in the fusion process still remain to be defined, myomaker appears to confer fusion competence to both of the fusing cells through the mixing of membrane lipids during the initial stage of membrane contact. Myomerger may be needed on only one of the fusing cells and it plays a role at a somewhat later stage, namely, the formation of a fusion pore. These two molecules are both necessary and also sufficient to permit the fusion of fibroblasts, which normally do not fuse with any other cell type.

Myostatin and the control of muscle growth

Growth involves not only proliferation and then an increase in size of muscle fibers, but also the ultimate cessation of these processes. How the growth of muscle is brought to a close was largely a mystery until the genetic basis for a mutant of cattle that causes a "doubled muscle" phenotype was uncovered (Fig. 3.7A). Some mutant muscles are over 50% larger than their normal counterparts. A similar mutant phenotype occurs in sheep, pigs, goats, horses, and even a human mutant has been reported (Fig. 3.7B). The defective gene product proved to be a protein, which was later called **myostatin**. Produced by muscle itself, myostatin is a growth factor, belonging to the large **transforming growth factor-β** family, and its main functions appear to occur during prenatal muscle development. Myostatin levels are higher in Type IIb (glycolytic) than in Type I (oxidative) muscle fibers. Another growth factor, **Activin A**, also acts along with myostatin to inhibit muscle growth. Both of these are, in turn, inhibited by yet another growth factor, called **follistatin**, which allows muscle growth by inhibiting the two inhibitors. Once produced in muscle, myostatin then enters the general circulation, where it is bound to regulatory factors to produce a latent form. Only after these complexes are broken down, does myostatin become active as a regulator of muscle growth.

Much of what is known about the way in which myostatin controls muscle growth has come from analysis of mutant animals. In cattle, the effects of myostatin become apparent as early as the stage when primary muscle fibers are forming in the embryo (see Fig. 3.1). Myostatin regulates the amount of proliferation of both muscle precursor cells and myoblasts and effectively controls the number of muscle fibers that will form. Mutant cattle (and other large animals) contain significantly more muscle fibers than normal, but the cross-sectional areas of the individual muscle fibers are not greatly increased. In contrast, experimentally derived myostatin mutant mice show an increase in both the number and size of their muscle fibers. Double-muscled cattle have an increased proportion of Type II (fast glycolytic) muscle fibers. Functionally, this leads to decreased endurance in these animals. In humans, myostatin levels are considerably increased in a number of diseases and during the aging process. This leads to muscle wasting.

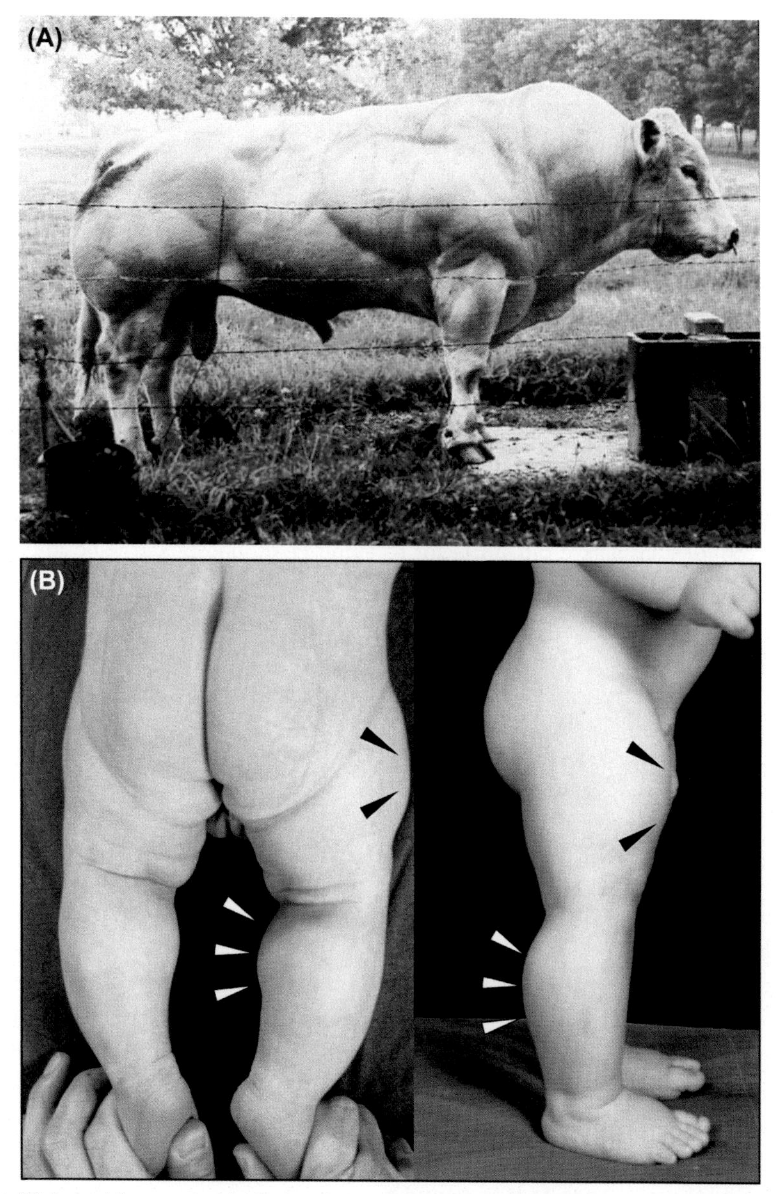

Figure 3.7 (A) A double-muscled bull, showing massive musculature caused by the lack of myostatin. (B) A case of double musculature in a human. Note the massive thigh musculature. *((A) From Carlson (2019), with permission. (B) From Schuelke et al. (2004), with permission.)*

Myostatin regulation of muscle growth is subject to control at many levels. All of these can result in different degrees of muscular hypertrophy or wasting. In addition to mutations of the DNA encoding the myostatin molecule itself, myostatin levels are

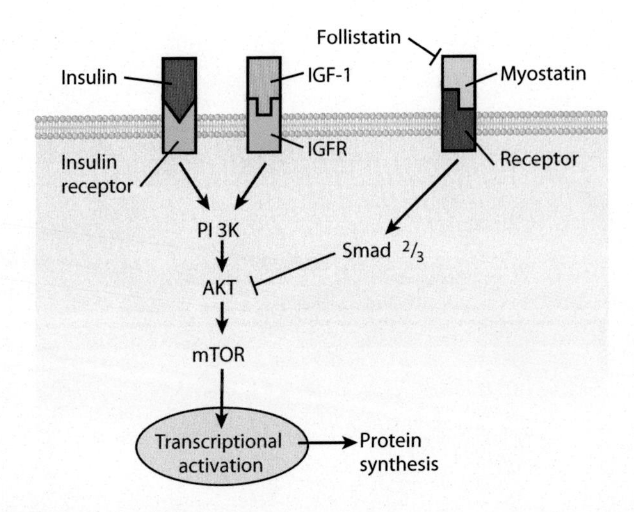

Figure 3.8 Simplified representation of the major protein synthetic pathway in muscle. *IGF*, insulinlike growth factor; *IGFR*, insulinlike growth factor receptor; *PI 3K*, phosphatidyloinositol 3-kinase; *TOR*, target of rapamycin.

regulated through a variety of forms of transcriptional modifiers (promoter or enhancer elements), posttranscriptional modifiers (e.g., microRNAs), epigenetic influences, and posttranslational regulators (proteins that bind to myostatin, e.g., follistatin). Myostatin also influences major metabolic pathways involved in muscle growth (see Fig. 3.8).

The intrauterine and early postnatal growth environment for muscle

Numerous factors influence muscle growth. Very important are mechanical factors—stretch, load, speed, and frequency of contractions. Equally important are metabolic and hormonal influences. At any given moment, all of these influences come into play, but before and shortly after birth the metabolic and hormonal environment is of particular importance in fostering the overall growth of muscle. During maturity, the balance shifts more toward mechanical influences, but without the appropriate metabolic and hormonal support, the full effect of mechanical influences might not be attained.

Metabolic pathways that foster growth

Nutrition and the metabolic pathways stimulated by ingested nutrients play a fundamental role in muscle growth, especially during the neonatal period, when muscle growth is most pronounced. In rats, for example, the proportion of muscle protein in relation to total body protein increases from 30% to 45% between birth and weaning.

Muscle growth is a function of the balance between protein synthesis and degradation. During rapid growth, protein synthesis greatly exceeds protein degradation, whereas under conditions of muscle loss or atrophy, protein degradation is the dominant process. Protein synthesis is the end product of a complex metabolic pathway that has both stimulatory and inhibitory inputs at different points along the way.

Protein synthetic pathways. A major stimulus for protein synthesis is the binding of **insulin** or **insulin-like growth factor-1** (IGF-1) with their appropriate receptors at the sarcolemma (Fig. 3.8). The changing molecular configuration of the bound receptor then activates a metabolic pathway that begins with **phosphatidylinositol 3-kinase** (PI 3K), which in turn activates **protein kinase B** (also known as **Akt**) as an intermediate step in the sequence. The final major link in this chain is **mTOR** (muscle target of **rapamycin**[1]), which is an important metabolic junction in the pathway leading to muscle protein synthesis. mTOR[2] is itself a kinase and functions as an integrator of a number of inputs, including the insulin/IGF-1 pathway described above. In addition to that pathway, other factors, such as amino acids, oxygen, nitric oxide, serum response factor (which influences satellite cell fusion), and testosterone, also feed into mTOR and can influence its activity. Integrating all of these influences, mTOR then influences the initiation of protein synthesis through its action in stimulating transcription of genes important for muscle growth. Knockout of the mTOR gene results in reduced postnatal growth, mainly because of a reduced size of fast muscle fibers. Growth of slow muscle fibers is considerably less affected. In contrast to the metabolically positive influences mentioned above, myostatin inhibits the mTOR pathway and inhibits muscle growth. Myostatin interacts with the Akt/mTOR pathway by decreasing Akt phosphorylation, thus reducing protein synthesis. In addition, it feeds into pathways leading into the activation of atrophy-inducing genes.

Protein degradation pathways. Even during periods of rapid growth, protein degradation is an essential function for maintaining the optimal condition of muscle. Contractile proteins can be damaged in normal muscle due to mechanical influences arising from muscle use or simply from random damage that occurs over time. In addition, whenever a muscle undergoes remodeling during early differentiation, growth, functional adaptation, or regeneration, protein degradation clears the way for the synthesis and reorganization of new arrays of contractile and other proteins. In fact, the half-life

[1] Rapamycin is a small molecule with potent antifungal activity that is derived from a soil bacterium discovered in an expedition to Easter Island (known also as Rapa Nui). mTOR was discovered during investigations of the mechanisms of action of rapamycin, which inhibits the functions of mTOR.

[2] Functionally, mTOR operates as a complex with other proteins. Two such complexes, **mTORC1** and **mTORC2**, have been identified. mTORC1 is the complex most heavily involved as a metabolic manager, and its major protein partner in that complex is called **raptor** (regulatory-associated protein of mTOR). mTORC1 is the complex that is sensitive to rapamycin, whereas mTORC2 is resistant to that antibiotic. Unless otherwise indicated, references to mTOR in this book refer to the mTORC1 complex.

of a myosin molecule in thick filaments of cultured myotubes has been estimated at between 3 and 24 h.

Protein degradation in muscle occurs through two dominant mechanisms, both of which also act in virtually almost all cells of the body. One, the **ubiquitin/proteasome pathway** (UPP), is the major system that cells use to degrade proteins. The other is **autophagy/lysosome-mediated proteolysis**, which assumes special importance in conditions of muscle wasting, such as disease, atrophy, and old age. In addition, glucocorticoid hormones can play a significant role in catabolic processes in muscle. These conditions are discussed in greater detail in Chapter 5.

Ubiquitin/proteasome pathway. Many muscle proteins destined for destruction are first tagged with a 76-residue protein, called **ubiquitin**, through the actions of **ubiquitin ligases** (E1—ubiquitin-activating; E2—ubiquitin-conjugating; E3—ubiquitin-ligating; Fig. 3.9). Then additional ubiquitin molecules are added to the first, forming a **polyubiquitin chain**. The ubiquinated proteins are then carried to **proteasomes** for actual degradation. Proteasomes, which are about half the size of ribosomes, consist of a cylindrical chamber composed of four stacked rings—each ring consisting of seven proteins (see Fig. 3.9). Regulatory proteins are attached to each end of the cylinder. Ubiquinated proteins attach to the regulatory elements at one end of the proteasome, and the ubiquitin chains are removed. Then the protein is unfolded, and the unfolded polypeptide chain passes through the core of the proteasome, where it is broken down

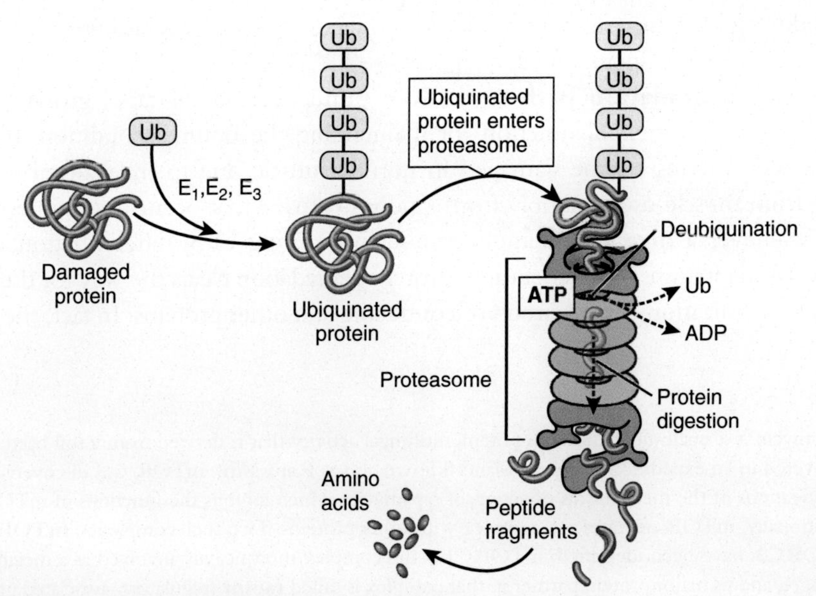

Figure 3.9 The ubiquitin proteasomal pathway (UPP) involved in the removal and breakdown of damaged proteins within a cell. *ADP*, adenosine diphosphatase; *ATP*, adenosine triphosphatase; *Ub*, ubiquitin.

into individual amino acids or small peptides. These are then available for reuse by the synthetic machinery of the cell.

Even in the earliest stages of myogenesis, proteasomes play a critical role in the orderly succession of myoregulatory factors. As satellite cells first become activated, Pax-3 and Pax-7, which maintain the satellite cell in the quiescent state, are removed through the UPP pathway. Then proteasomes remove a transcriptional inhibitory factor, **Id**, which then allows the synthesis of the myogenic regulatory factor, MyoD. As differentiation continues, MyoD is also removed by proteasomes. This process allows in succession the formation of Myf5, and later myogenin, as muscle fiber differentiation continues.

Autophagy/lysosome-mediated proteolysis. A second major pathway for removing damaged muscle proteins involves their being taken up by **lysosomes**—membrane-bound cytoplasmic vesicles that contain up to 60 degradative enzymes. This pathway of protein destruction assumes considerable prominence in a number of muscle diseases, which are characterized by either excessive destruction of proteins and other classes of molecules or incomplete breakdown of certain molecules and the accumulation of partial breakdown products within the lysosomes.

Systemic factors influencing intrauterine and early postnatal growth

The rate of muscle growth is by far the greatest in the late stages of pregnancy and the early postnatal period. In the fetus, muscle growth involves a mixture of myogenesis (formation of secondary muscle fibers) and an increase in both length and cross-sectional area of existing muscle fibers. Because the total number of muscle fibers is essentially set by the time of birth, postnatal muscle growth is confined to increasing the size of muscle fibers. Muscle fiber growth during this period is heavily dependent upon the nutritional status and hormonal environment of the individual. In utero, that environment is largely provided by the mother through the placental connection, through which maternal nutrients and hormones greatly influence the internal environment of the fetus. Postnatally, in addition to the nutritional status of the individual, activity levels come into play as modifiers of muscle growth.

Factors influencing intrauterine muscle growth

Under normal circumstances, the robust growth of muscle during the fetal period is based on a variety of factors—nutrient levels (especially amino acids), hormone (insulin) and growth factor (IGF-1) action, and oxygen supply. These feed into the Akt/mTOR pathway (see above) to stimulate the synthesis of muscle proteins, which in the fetus far exceeds degradative processes. During the earlier embryonic and fetal periods, this balance favors the production of new muscle fibers. As fetal development proceeds, the balance is tipped toward the growth of muscle fibers until by the late fetal period individual muscle fiber growth is the dominant factor underlying the overall increase in muscle mass.

Growth of fetal muscle is not an isolated event. The stem cells that give rise to myogenic cells also produce intramuscular adipocytes and later, fibroblasts. Understanding the formation of adipocytes is of vital importance in the meat sciences, because intramuscular fat formation provides the basis for the marbling of meat, which is a major determinant of its flavor. The generation of fibroblasts provides the cellular raw material for the formation of the various types of intramuscular connective tissue—the endo-, peri-, and epimysium. All of these processes are strongly influenced by the fetal environment, especially oxygen and nutrient levels. Much of our understanding of the basis for fetal muscle growth comes from studies of nutritional deficiency and placental insufficiency, on the one hand, and excessive nutrition on the other.

Undernourishment, in particular, demonstrates the importance of normal levels of protein synthesis on fetal muscle growth. Studies on a number of species of animals have shown that undernutrition during the embryonic periods when muscle fibers are forming results in decreases of up to 25%–30% from control levels of muscle fibers. Numbers of Type II muscle fibers are more severely impacted than those of Type I fibers. In addition, the numbers of satellite cells are adversely affected, with the result that by birth the numbers of satellite cells are reduced by as much as 33%. This not only impacts early postnatal muscle growth but also the ability of the muscle to regenerate later in life.

Undernutrition later in fetal life affects principally the growth of individual muscle fibers rather than their numbers, which are already mostly set by then. Under normal nutritional conditions, the actions of insulin, IGF-1, and even individual amino acids, especially leucine, stimulate the normal metabolic pathways to produce new muscle proteins well in excess of their destruction. Genetic knockouts of these growth factors result in a reduced musculature due to decreases in both muscle fiber number and size. Undernutrition and conditions of energy restriction result in a decrease of protein synthesis due to activation of the **tuberus sclerosis complex** and suppression of the mTOR metabolic pathway. Nutritional deficiency during pregnancy favors the formation of connective tissue over muscle fibers, a condition that not only affects overall function of the muscular system, but also the quality of meat in domestic animals. The balance between formation of muscle fibers and other tissues is also demonstrated with animals possessing mutations of the myostatin gene. For example, relative to normal, double muscled cattle (myostatin mutant) have not only much more muscle than normal (20% more in cattle and 2–3 times more in mice) but also less fat within the muscle.

After fetal nutritional deficiency, a period of postnatal catch-up growth is a well-documented phenomenon. Despite the catch-up response, certain effects are not normally overcome. Even in humans, a deficiency in fetal muscle growth typically persists throughout life, resulting in a reduced muscle mass. This would especially be a consequence of a reduced number of muscle fibers being formed. In addition, there is often an increased deposition of fat at the expense of muscle tissue. Equally important, reduced musculature and increased fat has important metabolic consequences into adulthood, especially a correlation between low muscle mass and future insulin resistance, resulting in the development of type II diabetes.

In domestic animals, overnutrition or administration of growth-promoting molecules to pregnant females also results in consequences to the musculature. Infusion of insulin or IGF-1 to pregnant sheep results in increased protein synthesis in the fetal muscles and an increase in overall growth. Overnourishment of pregnant sheep or pigs results in increased intramuscular adipogenesis and increased overall fatness at birth of their young.

Another important determinant of intrauterine muscle growth is the function of the placenta, which represents the conduit between maternal nutrition and nutrition of the fetus. **Placental insufficiency** involves both undernourishment and a hypoxic environment. Under conditions of placental insufficiency, blood flow and nutrient supply are diverted to support the development of vital organs, such as the brain, heart, and liver, at the expense of skeletal muscle, which can be reduced in mass by as much as 25%—40% from normal. Levels of both insulin and IGF, both of which support muscle growth, are reduced during placental insufficiency, and concentrations of norepinephrine and cortisol, which stimulate catabolic functions, are increased. As a result of the redistribution of blood flow from the peripheral to the central vasculature (basically supplying the brain and heart), both the number of muscle fibers and the number of myonuclei per fiber are reduced in placental insufficiency.

Factors influencing postnatal muscle growth

Postnatal muscle growth is extremely rapid during the immediate perinatal period, but like other mammalian growth processes, the rate of growth slows until at some point (after adolescence in humans) it effectively ceases. In mammals, the cessation of skeletal growth due to fusion of the epiphyseal plates in long bones marks the end of the normal growth period. Changes in muscle after that time can be viewed as adaptations to a changing environment. These changes are covered in subsequent chapters.

At birth, muscle development in humans and most other mammals ceases to involve the formation of new muscle fibers. Therefore, postnatal muscle growth consists principally of elongation and an increase in the cross-sectional area of the existing muscle fibers. Both of these processes require considerable protein synthesis, and for that an increase in the number of myonuclei is necessary to support this process. The source of these new myonuclei is satellite cells. In a number of newborn laboratory animals, satellite cells constitute almost one-third of the total number of nuclei located inside the basal lamina of a muscle fiber. During the period of rapid perinatal growth, the number of satellite cells decreases in proportion to the rate of overall body growth until at the succession of overall growth the percentage of satellite cells to myonuclei stabilizes at 4%—5% in many animals.

As in fetal development, the postnatal growth of skeletal muscle is dependent upon the metabolic pathways that drive protein synthesis, especially the Akt/mTOR pathway, and the hormonal influences (insulin and IGF-1) that modulate them. In addition, changes in the hormonal environment during adolescence, especially the increase in testosterone levels, have a significant effect on postnatal muscle growth.

Both **growth hormone** and **testosterone** have been implicated as enhancers of muscle growth. In the case of growth hormone, the effect appears to be a more general one that involves the entire musculoskeletal system. Although growth hormone enhances the IGF-1 regulation of the metabolic pathways leading to protein synthesis in both muscle and other tissues, it seems to play a particularly important in the formation of connective tissue—also an important component of muscle in the form of the various "-mysial" sheaths that enwrap anything from individual muscle fibers to entire muscles.

Testosterone has been shown to have a more direct effect on muscle growth. At the cellular level, this hormone has a stimulatory effect on the proliferation of satellite cells. Under the influence of testosterone, muscle fibers undergo hypertrophy and contain increased numbers of satellite cell-derived myonuclei. At the metabolic level, it both enhances the synthesis of muscle proteins and reduces their breakdown. Testosterone increases the local production of IGF-1 in muscle and also increases the intracellular utilization of amino acids during the production of muscle proteins. It decreases muscle protein breakdown by reducing the activity of the ubiquitin-proteosome pathway and possibly by interfering with glucocorticoid influences on protein breakdown pathways. In addition, testosterone, and possibly growth hormone as well, upregulates levels of follistatin, which inhibits myostatin, therefore reducing myostatin's inhibitory effects on muscle growth. Effects of testosterone on mature muscle are discussed in Chapter 4.

Nutritional levels, especially foods rich in amino acids, play a significant role in postnatal muscle growth. As is the case in the late fetus, nutritional deficiency in the perinatal period can selectively hinder muscle growth. On the other hand, good nutrition in postnatal individuals, both domestic animals and humans, allows an individual with retarded prenatal muscular growth to enter a "catch-up" phase and restore some of the deficient muscle mass. Unfortunately, much of the catch-up growth is diverted toward the formation of adipose tissue. In elderly human males and in sheep, low birth weight has been shown to result in significantly less muscle and a higher adipose tissue/muscle ratio than in a population with normal birth weights.

Physical activity and other mechanical factors also play an important role in early postnatal muscle growth. As is the case with the fetus, any growth of the skeleton stretches the attached muscles, causing them to add sarcomeres and grow in length to keep up with the skeletal growth. Not only physical activity, but even the increased mass of the growing body adds load to the musculature, with the resulting increase in cross-sectional areas of the muscle fibers and the muscles as a whole. This topic is discussed at greater length in Chapter 4.

References

Abmayr SM, Pavlath GK. Myoblast fusion: lessons from flies and mice. Development 2012;139:641–56.

Bell RAV, Al-Khalaf M, Megeney LA. The beneficial role of proteolysis in skeletal muscle growth and stress adaptation. Skel Muscle 2016;6:16. BioMed Central.

Brown LD, Hay Jr WW. Impact of placental insufficiency on fetal skeletal muscle growth. Mol Cell Endocrinol 2016;435:69–77.

Brown LD. Endocrine regulation of fetal skeletal muscle growth: impact on future metabolic health. J Endocrinol 2014;221:R13–29.

Chen Y, Zajac JD, MacLean HE. Androgen regulation of satellite cell function. J Endocrinol 2005;186: 21–31.

Cornelison DDW. Known unknowns": current questions in muscle satellite cell biology. Curr Top Dev Biol 2018;126:205–33.

Davis TA, Fiorotto ML. Regulation of muscle growth in neonates. Curr Opin Clin Nutr Metab Care 2009; 112:78–85.

Dayanidhi S, Lieber RL. Skeletal muscle satellite cells: mediators of muscle growth during development and implications for developmental disorders. Muscle Nerve 2014;50:723–32.

Du M, Tong J, Zhao J, Underwood KR, Zhu M, Ford SP, Nathanielsz PW. Fetal programming of skeletal muscle development in ruminant animals. J Anim Sci 2010;88(E. Suppl):E51–60.

Dumont NA, Wang YX, Rudnicki MA. Intrinsic and extrinsic mechanisms regulating satellite cell function. Development 2015;142:1572–81.

Hindi SM, Tajrishi MM, Kumar A. Signaling mechanisms in mammalian myoblast fusion. Sci Signal 2013; 6(272):re2.

Lehka L, Redowicz MJ. Mechanisms regulating myoblast fusion: a multilevel interplay. Semin Cell Dev Biol 2020;104:81–92.

Piccirillo R, Demontis F, Perrimon N, Goldberg AL. Mechanisms of muscle growth and atrophy in mammals and *Drosophila*. Dev Dynam 2013;243:201–15.

Rhoads RP, Baumgard LH, El-Kadi SW, Zhao LD. Roles for insulin-supported skeletal muscle growth. J Anim Sci 2016;94:1791–802.

Rodriguez J, Vernus B, Cheelh I, Cassar-Malek I, Gabillard JC, Sassi AH, Seiliez I, Picard B, Bonnieu A. Myostatin and the skeletal muscle atrophy and hypertrophy signaling pathways. Cell Mol Life Sci 2014;71:4361–71.

Schiaffino S, Reggiani C. Molecular diversity of myofibrillar proteins: gene regulation and functional significance. Physiol Rev 1996;76:371–423.

Schiaffino S, Dyar KA, Ciciliot S, Blaauw B, Sandri M. Mechanisms regulating skeletal muscle growth and atrophy. FEBS J 2013;280:4294–314.

Shan T, Xu Z, Wu W, Liu J, Wang Y. Roles of Notch 1 signaling in regulating satellite cell fates choices and postnatal skeletal myogenesis. J Cell Physiol 2017;232:2964–7.

Sharma M, McFarlane C, Kambadur R, Kukreti H, Bonala S, Srinivasan S. Myostatin: expanding horizons. IUBMB Life 2015;67:589–600.

West DWD, Phillips SM. Anabolic processes in human skeletal muscle: restoring the identities of growth hormone and testosterone. Physician Sportsmed 2010;38:97–104.

Wigmore PM, Evans DJR. Molecular and cellular mechanisms involved in the generation of fiber diversity during myogenesis. Int Rev Cytol 2002;216:175–232.

Yin H, Price F, Rudnicki MA. Satellite cells and the muscle stem cell niche. Phys Rev 2013;93:23–67.

CHAPTER 4

Muscle adaptation to increased use

In mammals, overall growth ceases when physical maturity is reached,[1] but this does not mean that the body or its components remain static. Virtually every part of the body is capable of adapting to changing circumstances, and skeletal muscle is no exception. Under an increased workload, a muscle gets larger, and when the workload diminishes, the muscle becomes correspondingly smaller. Muscles also adapt to their resting length. If a muscle is maintained in a lengthened position, it becomes longer, whereas one maintained in a shortened position reacts by becoming shorter.

Adaptations of muscles to changes in their mechanical environment do not affect only muscle fibers. As an integral part of a muscle, the connective tissue also adapts, but often in different ways. Similarly, the vascular supply to a muscle quickly adapts to changes in its patterns of use. All of these changes at the tissue level work in concert to allow a muscle to function optimally in meeting the functional demands imposed upon it. The overall nutritional state of the body plays a significant role in determining the extent to which muscles can adapt to changing circumstances.

Muscle adaptation is commonly viewed as responses to exercise regimes, which are commonly classified as either strength-building or endurance. Although adaptations of muscle occur in everyday life, these adaptive responses are accentuated through serious training protocols. As a result, most research on muscle adaptation has been focused on responses of both humans and laboratory animals to exercise training. Other than during growth, muscle adaptation to lengthening or shortening is less commonly seen in humans. Therefore, research on the responses of muscles to changes in resting length has been conducted principally on animal models.

Another form of adaptation is metabolic. Although a fundamental component of the response to endurance exercise, metabolic adaptation even prior to exercise also occurs in some nonmammalian species. A prime example is a species of sandpiper, which greatly increases the oxidative enzymatic activity in its flight muscles before undertaking a very long transoceanic migratory flight. These birds do so just before their flight by gorging on crustaceans that contain large amounts of polyunsaturated fatty acids. In this case, the adaptive change in muscle occurs in response to diet, rather than exercise.

[1] Animals, such as mammals and birds, that attain a maximum size upon reaching maturity are said to exhibit *determinate growth*. In contrast, fish, amphibians and reptiles, for example, continue to grow throughout life, although at a reduced rate. This is called *indeterminate growth*.

Muscle Biology
ISBN 978-0-12-820278-4, https://doi.org/10.1016/B978-0-12-820278-4.00005-0

Muscle adaptations to increased load (strength training) or increased duration of use (endurance training)

Muscles respond in dramatically different ways to increased loading (strength training) versus prolonged use (endurance training). Under conditions of increased load, especially after a regimen of strength training, the affected muscles undergo **hypertrophy**—an increase in cross-sectional area. On the other hand, the same muscles, if subjected to repeated use but without increased load, do not become significantly larger (Table 4.1). Instead, they become capable of many repeated contractions with much less fatigue than would have been the case before endurance training. Even at the molecular level, muscle fibers respond differently to endurance versus resistance training. Many of the intracellular changes that occur in response to endurance training are ultimately controlled at the level of transcription of genes, especially those related to mitochondria. On the other hand, many of the molecular controls that operate in response to resistance training take place at the posttranscriptional level, especially involving the synthesis of contractile proteins.

Although experiments involving human training programs have produced the most immediately relevant data, laboratory model systems have also yielded a wealth of information. One of the most commonly used models for producing hypertrophy is a synergist ablation model in rodents, in which all of the main postural muscles except one (usually the plantaris) are removed, and the animal must support its weight on the one remaining muscle. Another laboratory model is chronic electrical stimulation of a muscle. This model is coming into increasing use in the clinical treatment of atrophic muscle (see Chapter 5).

Table 4.1 Adaptations of muscles to resistance and endurance exercise.

	Resistance exercise	Endurance exercise
Muscle strength	+++	±
Muscle hypertrophy	++	±
Muscle fiber X.S. area	++	±
No. of myonuclei	++	±
Satellite cell %age	++	+
Contractile protein synthesis	++	+
Myoglobin synthesis	±	++
Mitochondrial volume	+	+++
Oxidative enzymes	±	+++
Glycolytic enzymes	++	+
Muscle glycogen use	++	±
Capillary density	±	++
Endurance	±	+++

Adaptations to increased load (hypertrophy)

Controlled exercise studies have clearly shown that a muscle that has been subjected to increased load (Box 4.1) becomes larger (undergoes **hypertrophy**). Not only does the entire muscle become larger, but the individual muscle fibers also increase in cross-sectional area. After resistance exercise, men show a slightly greater increase in the cross-sectional area of muscles than do women, although women show a slightly greater percentage increase in strength.

Some research has reported that new muscle fibers can form after increased load-bearing, especially in birds, but there seems to be little doubt that the main basis for muscular hypertrophy in humans is an increase in size of existing muscle fibers. Muscle fibers have been shown to split or branch after intensive load-bearing. This might be the result of regeneration of muscle fibers that were damaged from the exercise. Muscle fiber splitting may account for a slight increase in numbers of muscle fibers during hypertrophy. Changes in the connective tissue (extracellular matrix) also account for a small part of the increase in size of the hypertrophic muscle.

Studies on humans engaged in resistance training programs suggest that the initial phase of muscle enlargement soon after beginning a regimen of resistance training is due more to **edema** resulting from muscle damage than to an increase in muscle protein, but over the course of four to almost 20 resistance training sessions the increase in cross-sectional area of a muscle is increasingly due to the accretion of bundles of newly synthesized contractile proteins (Fig. 4.1).

A hypertrophic muscle in a normal individual is stronger—that is, it can generate greater force than it could before it became hypertrophic. Within days of beginning an intensive strength training program, a person is capable of exerting greater muscular force even before the muscle fibers have had a chance to become hypertrophic. This

BOX 4.1 Types of muscle contractions during exercise

Although the molecular mechanism of producing force (contraction) by a muscle is consistent, the conditions under which force is produced can vary in several ways. Specific terms are applied to these conditions. When a muscle shortens while contracting, for example, when doing a biceps curl, that contraction is called a **concentric contraction**. If a muscle contracts, but is unable to shorten because the load is too great, that contraction is called an **isometric contraction**. When a muscle elongates while contracting, which happens to leg muscles while walking or running downhill, the contraction is called an **eccentric contraction**.

All of these types of contractions can result in muscle hypertrophy if the load is great enough, but their effect on the muscle is not the same. Eccentric contractions, in particular, can cause mechanical damage to a muscle, especially in an untrained individual. After a period of training, muscle damage after eccentric contractions is considerably reduced.

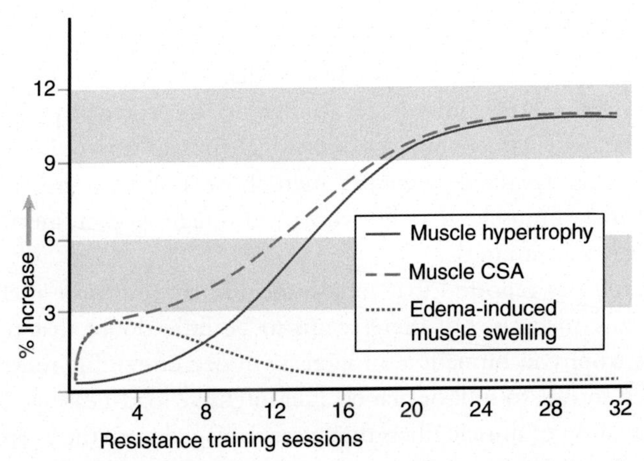

Figure 4.1 The relationship between edema and actual muscle hypertrophy as they contribute to total muscle cross-sectional area.

change is commonly attributed to **neural adaptation**. The mechanisms underlying neural adaptation remain to be well-defined, but increased motor unit firing and better coordination of movement have often been cited as components of neural adaptation.

Stimuli for muscular hypertrophy. There is probably no single stimulus that initiates muscular hypertrophy, but four elements of a high intensity training environment are commonly viewed as contributory factors. One is mechanical tension itself, which acts through still poorly defined mechanoreceptor mechanisms, and via pathways leading to increased protein synthesis. A second is actual damage to muscle fibers. Such damage can be at the level of some of the large macromolecules found within the muscle fiber, or it could extend to disruption of the muscle fiber itself, leading to degeneration and subsequent regeneration of new and presumably larger muscle fibers. A third likely stimulus is metabolic stress. This begins with the transient local hypoxia that occurs when muscles are contracting maximally. It is then translated into the build-up of metabolic byproducts and a resultant acidification of the environment. A fourth is the actions of certain hormones and growth factors, whose concentrations in the blood rise soon after a bout of resistance exercise.

The role of mechanical tension. Our understanding of how mechanical forces are translated into molecular events is still in its infancy, but many of the components required for such translation are known to be present in skeletal muscle fibers. Connections between the collagen and laminin molecules of extracellular matrix through the **integrins** and the **dystroglycan complex** that are situated in and beneath the sarcolemma are well established in skeletal muscle (Fig. 4.2). Another more recently recognized molecular complex (**Linker of Nucleoskeleton and Cytoskeleton** complex [LINC]) is now recognized to exist within the muscle fiber. The LINC connects

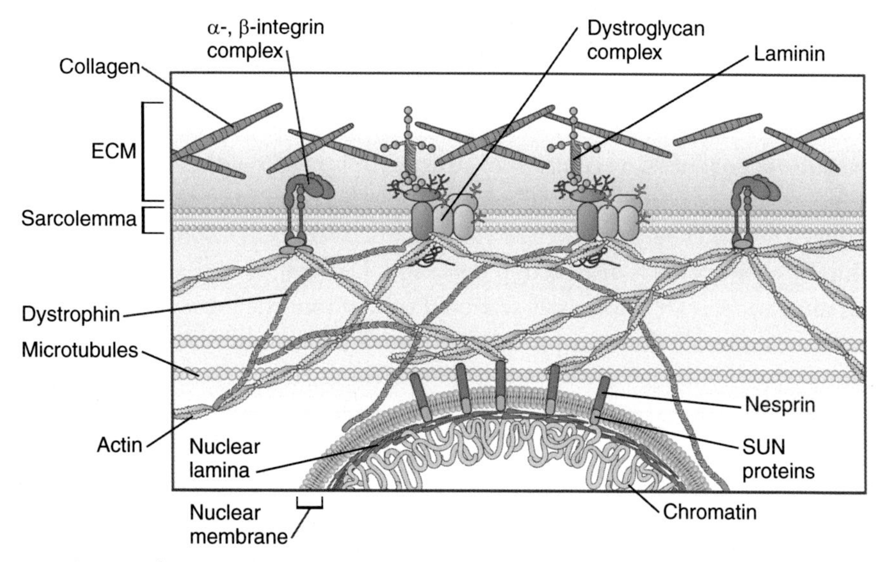

Figure 4.2 Structural connections between the connective tissue components surrounding a muscle fiber and the nucleus where they can influence gene expression.

the subsarcolemmal dystroglycan complex to the myonuclei and may be heavily involved in the proper spacing of myonuclei along the muscle fiber as well as providing a conduit between external mechanical forces and the regulation of gene expression in muscle fibers. One component of the LINC complex, the **nesprin** proteins, connects cytoskeletal elements to the outer nuclear membrane. The next link in the LINC chain comprises the **SUN** proteins, which connect the inner nuclear membrane to the **nuclear lamina**. The nuclear lamina is a meshwork of intermediate filaments, consisting mainly of A- and B-lamin proteins. The **nuclear lamins** bind to both transcription factors and to chromatin, and through this binding, they are able to influence gene expression. Genetic deletion of components of the LINC has been shown to be highly detrimental to muscle adaptation.

The transmission of mechanical forces from the extracellular environment to the nucleus through the connections outlined above can strongly influence intranuclear events, such as gene expression, chromatin organization, and the function of transcription factors, thereby influencing the production of contractile proteins. Some of the cellular functions that are strongly influenced by mechanosensitive pathways are ribosomal biogenesis—a crucial limiter of contractile protein synthesis, mTOR regulation of overall metabolic activity—including contractile protein synthesis, and several other less well investigated regulatory systems (mitogen-activated protein kinase [MAPK] and the Wnt/β-catenin pathway) that may play significant roles in the hypertrophic process.

The role of muscle fiber damage and regeneration in muscle hypertrophy. Many studies have shown that muscle fibers, especially fast muscle fibers, can become damaged during contraction under a heavy load, especially during eccentric contractions. Strong indicators of muscle damage are rapid increases of **creatine phosphokinase** and **myoglobin** in the blood. These leak out through the damaged sarcolemma of muscle fibers and enter the bloodstream. Another indicator of early muscle damage is **delayed onset muscle soreness**, a symptom of damage to both the muscle fibers and their surrounding connective tissue especially after a bout of eccentric exercise in an untrained individual.

Muscle regeneration is discussed in detail in Chapter 6. As a result of damage, regenerating muscle fibers are commonly scattered throughout the muscle. The role of regenerating muscle fibers in hypertrophy is not yet well understood, although there are some indications that myonuclei derived from satellite cells, as are those in regenerating muscle fibers, may not be as capable of a hypertrophic response as myonuclei from nonregenerating fibers.

Muscle fibers can also suffer micro-damage, specifically membrane disruption, especially after bouts of eccentric exercise. Within seconds, Ca^{++} accumulates at the site of damage. This is followed by the recruitment of several resealing proteins, such as **annexins**, which form a cap over the area of membrane damage. Formation of the initial repair cap, which involves the participation of nearby mitochondria, is completed within ten seconds of the injury. Further accumulation of proteins results in the formation of "scars" during the first 24 hours post-injury. Production of the scarring proteins involves migration of myonuclei toward the region of damage and their production of the proteins that make up the scars. As the micro-wound is being sealed, both the annexin proteins and mitochondria begin to sequester any excess Ca^{++} that has accumulated at the site of membrane damage. All of this repair activity occurs without the participation of satellite cells. Further research is necessary to determine the relative importance of macro- vs. micro-damage to muscle fibers after various forms of exercise.

Metabolic basis for muscle hypertrophy. As during growth, the size of a muscle is a reflection of the balance between protein synthesis and protein degradation. In hypertrophy, however, increased protein synthesis plays a much more prominent role than decreased degradation. The common metabolic meeting point for a variety of stimuli is **mTOR** (see p. 69), which serves as the molecular integrating center for a variety of stimulatory pathways that can begin with mechanical stimulation, growth factor activation or high amino acid concentrations (Fig. 4.3).

Two specific molecules directly play a role in mTOR activation. One is **phosphatidic acid**, a glycerophospholipid that may be one of the main mediators between mechanoreceptors within the sarcolemma and mTOR. The other is **Rheb**, a small member of the Ras superfamily of small GTPases (guanosine triphosphatases) that, when bound to GTP, activates mTOR (see Fig. 4.3). The action of Rheb is inhibited by another molecular mediator, **tuberous sclerosis complex 2** (Tsc 2). In order for Rheb to activate mTOR, the inhibitory action of Tsc 2 on Rheb itself must be inhibited. This is accomplished by the phosphorylation of Tsc 2, which then allows Rheb to activate mTOR. Several pathways leading from mechanoreceptors or hormone/growth factor receptors

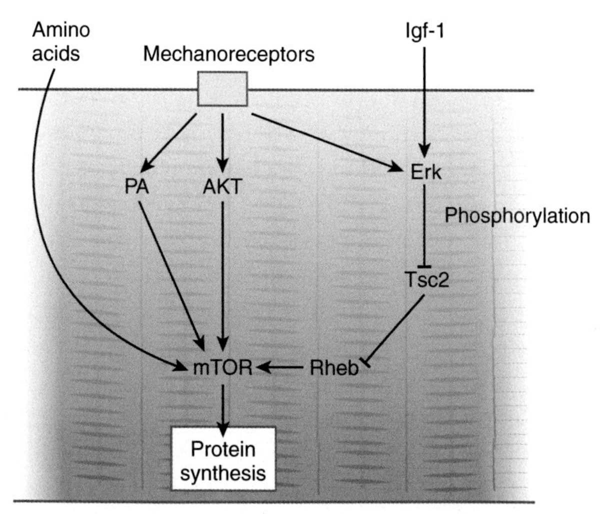

Figure 4.3 Exercise-related pathways leading to the synthesis of contractile proteins in muscle. *AKT*, protein kinase B; *IGF*, insulinlike growth factor; *PA*, phosphatidic acid; *TOR*, target of rapamycin; *Tsc*, tuberosclerosis complex.

can result in the phosphorylation of Tsc 2. The net result of both phosphatidic acid and Rheb action is the activation of mTOR, which in turn activates protein synthetic pathways, in part through increasing numbers of ribosomes within the muscle fibers.

A high concentration of amino acids, especially **leucine** and other essential amino acids, is also stimulatory to protein synthesis in resistance exercise-stimulated muscles. This is accomplished by several means, including translocating mTOR closer to Rheb within the sarcoplasm. Resistance exercise sensitizes muscle fibers to the stimulatory effects of amino acids as long as 24–48 h after a bout of exercise.

Hormonal and growth factor responses to muscle loading. At least three hormones or growth factors may play a role in muscular hypertrophy following intense resistance exercise. The most prominent is **testosterone**, a steroid hormone mainly produced in the Leydig cells in the testes of men, but also a product of the adrenal cortex and some local tissues. Although there are considerable discrepancies in the literature, a general finding is that in adult males circulating testosterone increases within minutes of heavy resistance exercise, but then returns to close to baseline levels within 30 min. Such an acute postexercise increase does not occur in women or in boys or older men.

Testosterone is an **anabolic steroid**, which within a muscle fiber binds to a cytoplasmic **androgen** receptor and significantly increases protein synthesis through its action on the transcription of genes that code for muscle proteins. Testosterone also functions as an anticatabolic agent by inhibiting the degradation of proteins by blocking a cortisol receptor (**cortisol**, an adrenal cortical steroid hormone, stimulates the breakdown of muscle proteins). The net result of these functions is an increase in total muscle contractile protein and consequently larger muscle fibers. Although testosterone exerts a demonstrable effect on mature muscle, its anabolic effects are considerably stronger on growing muscle.

Like testosterone, **growth hormone** (a pituitary hormone) levels also increase within the first 30 min following intense resistance exercise and then fall to resting levels. A major stimulus for growth hormone release appears to be the hypoxia and lowered pH that accompanies heavy resistance exercise. Growth hormone is a synergist of testosterone and may be an important factor involved in the anabolic effects of testosterone.

Insulin-like growth factor-1 (IGF-1) is a powerful growth factor involved in normal muscle growth (see p. 72). Released by the liver and also produced locally as a muscle-specific form, it stimulates protein synthesis and free fatty acid utilization, as well as increasing the sensitivity of cells to insulin. Although measurements of blood levels of IGF-1 postexercise have given varying results, there is a consensus that levels of muscle-derived IGF-1 increase after exercise.

Recent research has revealed an unexpected role of myogenic regulatory factors (MRFs) in regulating the size of adult muscle fibers. MRFs, which are transcription factors, play vital roles in embryonic myogenesis (see Chapter 2), but for years they were assumed to cease functioning once muscle fibers had matured. One of the MRFs, **MRF4**, is expressed in skeletal muscle (but not in cardiac muscle) throughout postnatal development. RNA knockdown experiments have shown that in the absence of functional MRF4, postnatal muscle fibers undergo hypertrophy, and atrophy in denervated muscles is much reduced. These responses occur because another MRF, **myocyte enhancer-binding factor 2** (**MEF2**), is derepressed. MEF2 activity stimulates a broad program of muscle gene expression that leads to muscle fiber hypertrophy.

Increase in muscle fiber size. The most prominent component of muscle hypertrophy is an increase in the cross-sectional area of individual muscle fibers, especially Type II fibers, in response to heavy loads. Most of this increase is due to the addition of arrays of contractile proteins in the form of newly added myofibrils. According to the nuclear domain theory (see p. 63), a muscle fiber must add new nuclei in order to increase the amount of contractile proteins that it can produce. Because myonuclei are incapable of dividing, the new nuclei must come from outside the muscle fiber. The major source of new myonuclei is satellite cells, which divide in response to strenuous weight-bearing exercise. As was seen in the case of muscle fiber growth (see p. 64), many of the daughter cells resulting from satellite cell division fuse with the muscle fiber, and their nuclei become myonuclei, which then begin to produce the RNAs that begin the process of protein formation.

Although most studies on humans have shown around 30% increases in satellite cells after resistance exercise, whether or not nuclear contributions from satellite cells are obligatory for muscle fiber hypertrophy remains an open question, with research reports supporting either side of the question. It does appear, however, that under some circumstances, mainly experimental, some degree of fiber hypertrophy can occur in the absence of incorporation of satellite cell nuclei. Increasing AKT/mTOR activity appears to alter overall synthetic activity to allow greater contractile protein synthesis without increasing the number of nuclei. Muscle fiber hypertrophy without a satellite cell contribution can also occur in myostatin mutants, but in this case the increase in muscle fiber cross-sectional area may be due to intracellular components other than contractile proteins. Overall, it appears that in the short term, muscle fiber hypertrophy can occur in the absence of satellite cells, but that for long-term hypertrophy to be maintained, satellite cells play a vital role.

The ability to experimentally knock out satellite cells from mouse muscle has provided new insight into the role of satellite cells in the hypertrophic process. Young growing mice require satellite cells for any significant degree of muscle fiber hypertrophy to occur. On the other hand, adult mice lacking in satellite cells are capable of a robust hypertrophic response. One explanation for this difference is that myonuclei in adults have a high transcriptional reserve that allows them to produce more contractile proteins in the absence of additional satellite cell-derived myonuclei, whereas in growing animals any transcriptional reserve is already taken up by the growth process.

Studies on humans have shown that muscle hypertrophy in response to increased load is less pronounced in elderly individuals than in young adults. Although many factors might contribute to this diminished response (see Chapter 8), one is a reduction in the response of satellite cells to the hypertrophic stimuli and a consequent reduction in the contribution of satellite cell nuclei to the muscle fibers. Not only does the number of satellite cells decrease during aging, but their activation is also reduced due to reductions in growth factors, such as IGF.

Changes in muscle fiber types. After prolonged periods of training, shifts in the populations of muscle fiber types occur. After resistance training, there is little change between proportions of Type I and II fibers. Within Type II fibers a small, but consistent percentage of Type IIx fibers becomes converted into Type IIa (fast-oxidative) fibers. Overall, with strength training, there is a slight shift toward slower, more oxidative muscle fibers.

Microvascular response to muscle contractions. The microvascular branches of the intramuscular arterial system (often called the **resistance vasculature**) become smaller with each branching point until the **terminal arterioles**, which then supply groups of 15–20 capillaries, forming what is known as a **microvascular unit**. With even a single contraction, the microcirculation responds by significantly increasing the blood flow through vasodilation to the region of the muscle that it supplies. The degree of increase is determined by the needs of a specific area, and it is almost instantaneous. Two main factors contribute to this increase. One is sympathetic relaxation of the smooth muscle layer in the arterioles. The second is a near-instantaneous upstream signal from the capillaries to the resistance arterioles. Stimulated by mechanical features of a single contraction and starting in the capillaries, endothelial cells initiate an electrical signal (hyperpolarization) that travels upstream—first via gap junctions between endothelial cells and slightly later involving a Ca^{++} wave. The net result is an increase in blood flow proportional to the needs of the muscle. Both the density of the microcirculation and its function respond proportionally to the requirements of exercised muscles.

Other changes in hypertrophic muscle. In contrast to muscles adapted for endurance exercise, changes at the cellular level in hypertrophying muscles, other than increases in contractile proteins, have been harder to document, with research studies often showing different effects with different modes of training. Changes in the overall capillary supply are not great after resistance exercise, but given a shift toward more Type IIa (fast oxidative/glycolytic) from Type IIx (fast glycolytic) muscle fibers, an increase in the capillary supply to these type-shifted fibers would be expected. Within hypertrophied muscle fibers, changes in the sarcoplasmic reticulum and Ca^{++} release mechanisms have been reported.

A relatively little investigated aspect of muscle hypertrophy is the effect of resistance exercise on the connective tissue of muscle. There is no doubt that increasing loads must be met by a remodeling and strengthening of the connective tissue and tendons of a muscle in order to prevent significant damage to the muscle. Increasing evidence suggests that in addition to their roles in supplying new myonuclei to muscle fibers, satellite cells also communicate with their nearby extracellular matrix through the release of growth factors and other influences that activate synthetic processes within the connective tissue of a muscle.

Adaptations to increased duration of use (endurance exercise)

Muscle responds to endurance exercise in dramatically different ways from its response to increased load. Neither whole muscles nor muscle fibers enlarge to a significant degree, and changes in strength (force) are minimal. In fact, some studies have shown that specific force (force/cross-sectional area) of muscle fibers may actually decrease after a period of endurance training. On the other hand, endurance-trained muscles become much less fatigable, thanks to changes at the tissue, cellular, and metabolic levels. Other changes occur at the whole body level, but these are beyond the scope of this book.

The dominant response to endurance training is a shift toward "slowness" in the population of muscle fibers and all that such a shift entails. The principal shift occurs within the Type II muscle fibers, with the conversion of glycolytic Type IIx fibers in humans (Type IIb in rodents) to oxidative/glycolytic Type IIa fibers. During the conversion, fast and slow myosin heavy chains may coexist within a single muscle fiber. An exercise-induced transformation of Type II to Type I muscle fibers is much less likely to occur in humans. The shift within Type II muscle fibers involves the replacement of the heavy chains of fast myosin by slow myosin, a large increase in mitochondrial volume and number, increased sensitivity to insulin, a metabolic shift from utilization of carbohydrates to lipids, and an increase in the production of internal antioxidants. Corresponding to these changes within the muscle fibers is a substantial increase in the capillary supply to the newly transformed slow muscle fibers.

Changes in contractile proteins. Responses to endurance training are mediated through a variety of molecular programs and metabolic pathways. The main pathway leading to changes in the synthesis of contractile proteins differs from that leading to mitochondrial adaptations to endurance exercise.

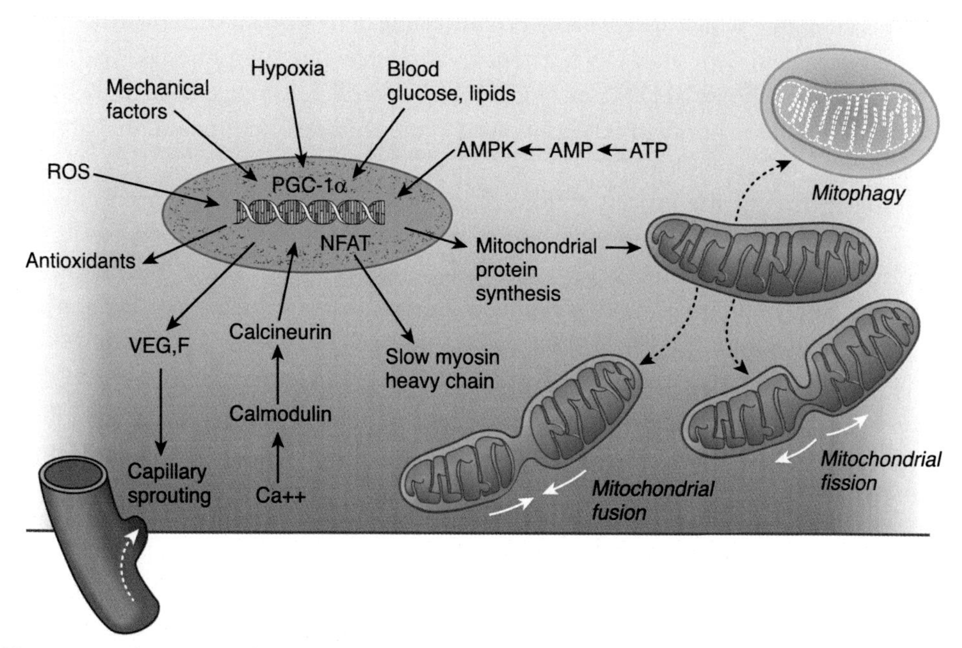

Figure 4.4 Adaptations of muscle fibers to endurance exercise. Arrows pointing into the nucleus at the left indicate factors that influence gene expression. Arrows leading out indicate effects on the muscle fiber. *AMP*, adenosine monophosphate; *AMPK*, adenosine monoposphate kinase; *ATP*, adenosine triphosphate; *NFAT*, nuclear factor of activated T-cells; *PGC*, peroxisome proliferator-activated receptorγ coactivator-1α; *ROS*, reactive oxygen species; *VEGF*, vascular endothelial growth factor.

The release of Ca^{++} during rhythmic muscle contractions begins a chain of events whereby the Ca^{++} ions are bound by the protein **calmodulin**, which activates **calcineurin** within the sarcoplasm of the muscle fiber. Calcineurin, in turn, enters the nucleus and activates transcription factors (e.g., NFAT [nuclear factor of activated T-cells]), which activate the transcription of genes encoding for slow muscle contractile proteins (Fig. 4.4).

Mitochondrial adaptations. One of the earliest recognized effects of endurance training was a significant increase in the volume (number, but mostly size) of mitochondria in the muscle fibers that underwent a glycolytic to oxidative/glycolytic transformation. The term applied to this increase is **mitochondrial biogenesis**. This increase in mitochondrial content is a structural reflection of a significant metabolic shift that results in more efficient ATP-based aerobic energy production by the muscle fibers.

Muscle fibers contain two subpopulations of mitochondria. One—**interfibrillar mitochondria**—is closely associated with the contractile proteins and provides immediate energy for cross-bridge formation and release during muscle fiber contractions. A second—**subsarcolemmal mitochondria**—is located, as the name implies, close to the sarcolemma and functions to provide energy for functions such as maintaining membrane potential and the transport substances across the sarcolemma.

Mitochondria, which are probably evolutionarily derived from bacteria that inhabited early cells, contain their own DNA (circular, rather than linear as is the case with nuclear DNA), but mitochondrial DNA encodes for only a small percentage of all mitochondrial proteins. The vast majority of RNA involved in the synthesis of mitochondrial proteins is derived from the myonuclei.[2]

Fundamental to the process of mitochondrial biogenesis resulting from endurance exercise is **PGC-1α** (peroxisome proliferator-activated receptorγ coactivator-1α). Similar to mTOR in another aspect of muscle function, PGC-1α acts as a master regulator for the genetic program that produces mitochondrial proteins (see Fig. 4.4). The synthesis of PGC-1α is promoted by endurance exercise, and it acts within both the myonuclei and mitochondria. Even a single bout of endurance exercise has been shown to increase levels of PGC-1α.

Mitochondrial biogenesis is a collective term that includes both **mitochondrial fission** (the splitting of a single mitochondrion) and fusion (the joining of individual mitochondria to form a single large mitochondrion). Another important component of the process of mitochondrial biogenesis is the removal of damaged mitochondria through an autophagic process, called **mitophagy**. In mitophagy, a damaged mitochondrion is ultimately fused with a lysosome. Lysosomal enzymes then break down the damaged mitochondrion into its component molecules, many of which are then recycled to form new mitochondria.

Metabolic adaptations. The endurance exercise-induced shift from an overall glycolytic to an oxidative metabolic profile has many components. Instead of relying on the anaerobic breakdown of intracellular glycogen—an energetic inefficient process—endurance-trained muscle fibers obtain the bulk of their energy-producing substrates from the blood. This is facilitated by adaptations that increase transport of these substances across the sarcolemma, such as greater sensitivity to insulin, which increases the transport of glucose in the blood into the muscle fiber. Even a single bout of exercise results in greater insulin sensitivity for hours or even days. Overall, for a given level of exercise, prior endurance training results in a relative reduction of carbohydrate (glycogen, glucose) utilization and a corresponding increase in the use of lipid as a metabolic substrate. These effects, however, are seen only in the specific muscles being trained and not in other muscles that are not involved in a particular training program. This shows that the basis for this adaptation to endurance exercise is mainly intrinsic to muscles, rather than a general systemic adaptation.

[2] The mitochondrial genome encodes for 13 mitochondrial membrane proteins, 22 tRNAs and 2 rRNAs. Over 1000 other mitochondrial proteins are synthesized from myonuclear-derived mRNAs in the sarcoplasm.

Reactive oxygen and antioxidants. Both resting and exercised muscle produces **reactive oxygen species (ROS)** (e.g., hydrogen peroxide and superoxides), which can be damaging to cells over a long period. One of the major theories of aging attributes cellular aging to exposure to reactive oxygen (and possibly nitrogen) species. In muscle, fast muscle fibers produce considerably more superoxides than do slow muscle fibers. The production of ROS increases with exercise, but endurance exercise also results in the production of antioxidants (e.g., catalase for hydrogen peroxide and superoxide dismutase for superoxides) by the same muscle fibers that produce ROS. Interestingly, ROS act on PGC-1α, which is then stimulated to initiate a chain of molecular events leading to the production of antioxidants directed against the oxidants that stimulated their production in the first place.

Changes in the microvasculature. One of the major adaptations of muscle to endurance exercise is an increase in the capillary supply to muscle fibers, especially those that have undergone a fast to slow transformation. Under the influence of PGC-1α, hypoxia, and mechanical factors, muscle fibers produce **VEGF** (vascular endothelial growth factor), which stimulates the local capillaries to produce endothelial sprouts, which lead to the formation of new microvascular channels. Even a single bout of exercise can elicit this response. The resulting increased capillary supply to the slow muscle fibers provides both a greater surface area for the exchange of metabolic substrates between the muscle fibers and the blood and a reduction in the velocity of blood flow, which provides more time for exchange to occur.

In addition to causing changes in the microvasculature, endurance exercise induces changes in larger blood vessels, as well. The arteries in the exercised muscles develop a larger lumen (cross-sectional area), and the thickness of the walls also decreases. The larger lumen permits greater blood flow during exercise.

Reversal of training gains

The ability of skeletal muscle to adapt to changing circumstances applies not only to training (see above), but to detraining, as well. The fundamental principle is that muscle quickly adapts to its present needs, so that if the endurance requirements or mechanical load are reduced, the size and functional capacities of the involved muscles become correspondingly reduced. Training and detraining changes strongly affect cardiovascular function, as well, but this is outside the scope of this book.

Intramuscular adaptations are particularly pronounced when endurance exercise in initiated or discontinued (Fig. 4.5). Most prominent are changes associated with mitochondrial function. Greater oxidative capacity occurs quickly after the onset of an endurance exercise program, but this capacity is lost at an even greater rate upon cessation of the program (Fig. 4.6). Recovery upon reinitiation of training takes longer than rate of loss.

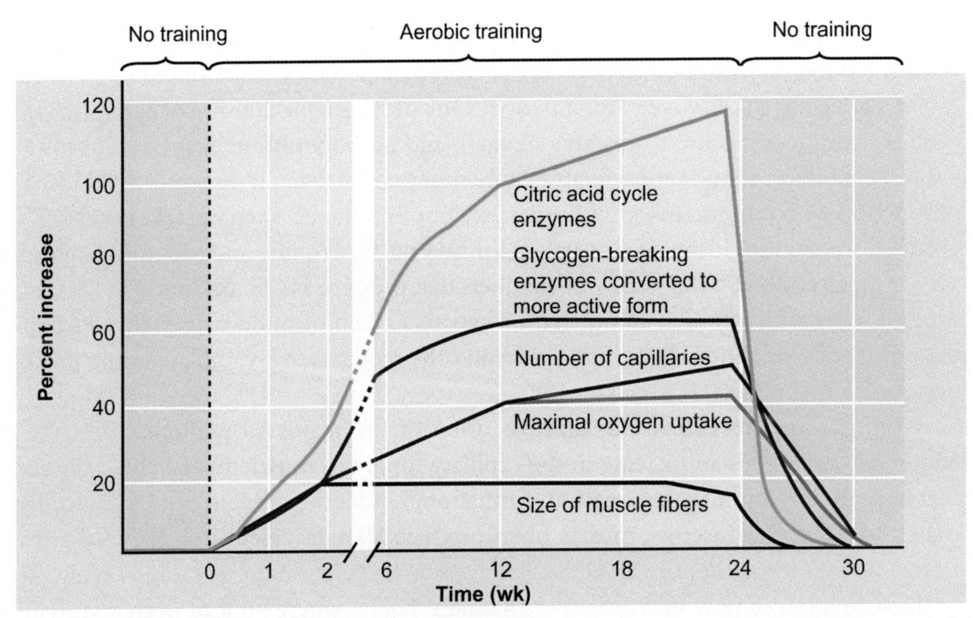

Figure 4.5 Effect of aerobic training and detraining on a number of properties of skeletal muscles. *(From Thibodeau and Patton (2007), with permission.)*

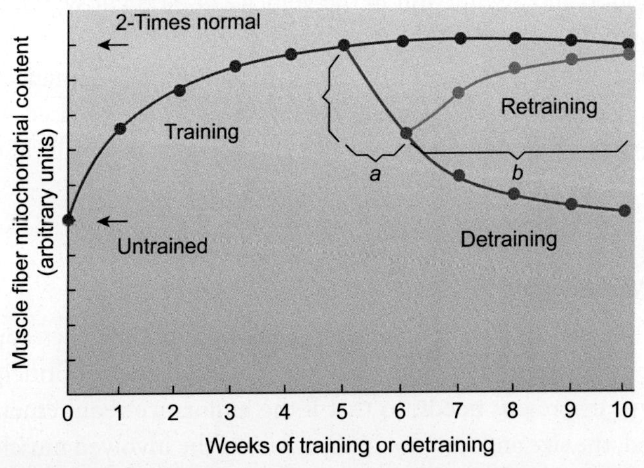

Figure 4.6 Effect of training and detraining on the mitochondrial content of skeletal muscle. After a week of detraining (*a*), almost 50% of the gain in mitochondrial content is lost. Upon retraining (*b*), it takes 4 weeks to recover what was lost in 1 week of detraining. *(From Powers and Howley (2018), with permission.)*

Upon the cessation of resistance training, the rate of loss of muscle function is considerably less than that following endurance training. Even after 6 months of detraining after a training program, less than half of the gain in strength is lost. Loss in cross-sectional area of muscle fibers is even less than that (in one study from 2% to 13% between Type I and Type IIx fibers). In addition, it is easier to maintain strength with minimal exercise after detraining from a resistance program than an endurance program.

Detraining represents one end of a spectrum of inactivity that can be imposed upon muscle. At the other end of the spectrum are bed rest, conditions of weightlessness, as seen with astronauts, and the complete loss of muscle function due to interruption of the motor nerve supply. These conditions lead to muscle atrophy, the subject of Chapter 5.

Adaptation to changes in length

In addition to adapting to changes in load or duration of exercise, skeletal muscle also reacts to changes in length by adjusting the number of sarcomeres and sarcomere length to levels that produce the optimum force, velocity, and power for the affected muscle. Changes in length can include either chronic lengthening or shortening.

Although the biology of adaptations to changes in muscle length has been most extensively studied in the laboratory, a number of real-life situations affecting humans make these studies relevant. The need for muscles to get longer during normal growth has already been discussed in Chapter 3. Certain situations in which limbs are immobilized in casts can result in muscles fixed in either lengthened or shortened positions for extended periods of time. An orthopedic procedure (Ilizarov method) is sometimes applied to elongate an extremity that is shorter than the other. This involves cutting through a long bone and then progressively extending the limb during the fracture-healing process. As the skeleton is lengthened, the muscles traversing the fracture site are also stretched. A classic case of muscle shortening involves the effect of wearing high heel shoes on the length of the gastrocnemius muscle.

Both muscle fibers and the associated connective tissue (including tendons) respond to stretch. Initially after stretch, existing sarcomeres are stretched beyond their optimal length. At a structural level, muscle fibers adapt by adding new sarcomeres at the ends of the muscle fibers in order to restore the most efficient length. In a laboratory model, stretched soleus muscles in rodents added four new sarcomeres per myofibril per hour, or 8000 sarcomeres per hour per muscle fiber.

In addition to structural changes, stretched muscle fibers also show prominent and rapid functional changes, which can involve both the contractile apparatus and its metabolic support system. Stretch, combined with electrical stimulation, results in rapid hypertrophy (a 35% increase in a week) of the fast rabbit tibialis anterior muscle. Since electrical stimulation alone does not result in hypertrophy, stretch alone appears to be a powerful influence. Such hypertrophy is accompanied by a more than twofold increase in total RNA (mainly ribosomal RNA), which supports increased synthesis of contractile proteins. Stretching stimulates the secretion of insulin-like growth factor-1 by the muscle fibers. In an autocrine manner, the secreted IGF-1 then acts back on the muscle fiber to stimulate greater protein synthesis. Chronic stretch of a fast muscle also produces a shift toward a greater percentage of slower muscle fibers, an indication that a change in the mechanical environment can produce a significant change in gene expression. Experience with muscle pain and limitation of movement after limb lengthening procedures strongly suggests that remodeling of the connective tissue within both the stretched muscles and may lag behind the lengthening of the muscle fibers. The adaptability of tendons and intramuscular connective tissue to changes in muscle length decreases with age.

Adaptation to a shortened position has been studied in animal models, with limbs casted in a shortened position, and has been modeled in humans who wear high-heeled shoes. With shortened sarcomeres, there is excessive overlap between thick and thin filaments. Sarcomere removal is the principal way to restore the conditions for most effective contractions of the shortened muscle. In the human high heel model, wearing 13 cm heels shortens the gastrocnemius muscle by 5%, whereas the lengths of the soleus muscle and Achilles tendon remain unchanged. The effects of shortening are not constant throughout the gastrocnemius muscle. Changes in sarcomere length are most pronounced midway in the muscle belly, and it is in this region where the most sarcomere loss occurs. Computer modeling suggests that up to 30% of the sarcomeres in this region could be lost. Upon returning to wearing flat-heeled shoes, sarcomeres and the connective tissues surrounding the muscle fibers in the central gastrocnemius are greatly stretched, often with resulting muscular pain.

In animal casting models, other changes besides sarcomere removal have been shown. Casting the soleus muscle of rodents in a shortened position results in a 40% decrease in the number of sarcomeres in series. In the slow soleus muscles of mice, which do not normally have Type II muscle fibers, the expression of fast myosin genes takes place as soon as 1 day after a limb is casted in a shortened position.

References

Abreu P, Mendes SVD, Ceccatto VM, Hirabara SM. Satellite cell activation induced by aerobic muscle adaptation in response to endurance exercise in humans and rodents. Life Sci 2017;170:33–40.

Altana V, Geretto M, Pulliero A. MicroRNAs and physical activity. MicroRNA 2015;4:74–85.

Blaauw B, Reggiani C. The role of satellite cells in muscle hypertrophy. J Muscle Res Cell Motil 2014;35:3–10.

Booth FW, Ruegsegger GN, Toedebusch RG, Yan Z. Endurance exercise and the regulation of skeletal muscle metabolism. Progr Mol Biol Transl Sci 2015;135:129–51.

Caiozzo VJ, Utkan A, Chou R, Khalafi A, Chandra H, Baker M, Rourke B, Adams G, Baldwin K, Green S. Effects of distraction on muscle length: mechanisms involved in sarcomerogenesis. Clin Orthop Relat Res 2002;403S:S133–45.

Davis SR, Wahlin-Jacobson S. Testosterone in women – the clinical significance. Lancet -Diabet/Endocrinol 2015;3:980–92.

Goldspink G, Williams P, Simpson H. Gene expression in response to muscle stretch. Clin Orthop Relat Res 2002;403S:S146–52.

Goldspink G. Changes in muscle mass and phenotype and the expression of autocrine and systemic growth factors by muscle in response to stretch and overload. J Anat 1999;194:323–34.

Goldspink G. Mechanical signals, IGF-1 gene splicing, and muscle adaptation. Physiology 2005;20:232–8.

Goodman CA. Role of mTORC1 in mechanically induced increases in translation and skeletal muscle mass. J Appl Physiol 2019;127:581−90.
Jacobs BL, Goodman CA, Hornberger TA. The mechanical activation of mTOR signaling: an emerging role for late endosome/lysosomal targeting. J Muscle Res Cell Motil 2014;35:11−21.
Kirby TJ. Mechanosensitive pathways controlling translation regulatory processes in skeletal muscle and implications for adaptation. J Appl Physiol 2019;127:608−18.
Kraemer WJ, Ratamess NA, Nindl BC. Recovery responses of testosterone, growth hormone, and IGF-1 after resistance exercise. J Appl Physiol 2017;122:549−58.
Marcotte GR, West DWD, Baar K. The molecular basis for load-induced skeletal muscle hypertrophy. Calcif Tissue Int 2015;96:196−210.
McGlory C, Phillips SM. Exercise and the regulation of skeletal muscle hypertrophy. Prog Mol Biol Transl Sci 2015;135:153−73.
Murach KA, Fry CS, Kirby TJ, Jackson JR, Lee JD, White SH, Dupont-Versteegden EE, McCarthy JJ, Peterson CA. Starring or supporting role? Satellite cells and skeletal muscle fiber size regulation. Physiology 2018;33:26−38.
Murlasits Z, Kneffel Z, Thalib L. The physiological effects of concurrent strength and endurance training sequence: a systematic review and meta-analysis. J Sports Sci 2018;36:1212−9.
Newsom SA, Schenk S. Interaction between lipid availability, endurance exercise and insulin sensitivity. Med Sport Sci 2014;60:62−70.
Otto A, Patel K. Signalling and the control of skeletal muscle size. Exp Cell Res 2010;316:3059−66.
Qaisar R, Bhaskaran S, van Remmen H. Muscle fiber type diversification during exercise and regeneration. Free Radic Biol Med 2016;98:56−67.
Roman W, Pinheiro H, Pimental MR, et al. Muscle repair after physiological damage relies on nuclear migration for cellular reconstruction. Science 2021;374:355−9.
Rowe GC, Safdar A, Arany Z. Running forward. New frontiers in endurance exercise biology. Circulation 2014;129:798−810.
Schiaffino S, Dyar KA, Calabria E. Skeletal mass is controlled by the MRF4-MEF2 axis. Curr Opin Clin Nutr Metab Care 2018;21:164−7.
Schoenfeld BJ. The mechanisms of muscle hypertrophy and their application to resistance training. J Strength Condit Res 2010;24:2857−72.
Terena SML, Fernandes KPS, Bussadori SK, Deana AM, Mesquita-Ferrari RA. Systematic review of the synergist muscle ablation model for comprehensive hypertrophy. Rev Assoc Med Bras 2017;63:164−72.
Tiogo M, Boutellier U. New fundamental resistance exercise determinants of molecular and cellular muscle adaptations. Eur J Appl Physiol 2006;97:643−63.
Vingren JL, Kraemer WJ, Ratamess NA, Anderson JM, Volek JS, Maresh CM. Testosterone physiology in resistance exercise and training. Sports Med 2010;40:1037−53.
Yan Z, Okutsu M, Akhtar YN, Lira VA. Regulation of exercise-induced fiber type transformation, mitochondrial biogenesis, and angiogenesis in skeletal muscle. J Appl Physiol 2011;110:264−74.
Zoellner AM, Abilez OJ, Boel M, Kuhl E. Stretching skeletal muscle: chronic muscle lengthening through sarcomerogenesis. PLoS One 2012;7:e45661.
Zoellner AM, Pok JM, McWalter EJ, Gold GE, Kuhl E. On high heels and short muscles: a multiscale model for sarcomere loss in the gastrocnemius muscle. J Theor Biol 2015;365:301−10.

CHAPTER 5

Muscle adaptation to decreased use

Adaptive properties of muscles are usually viewed as responses to increased use, especially increased load or duration of exercise. It is important to remember, however, that muscles adapt to almost any change in the functional or mechanical environment in which they operate. This includes decreased use. Upon cessation of exercise, the involved muscles surprisingly quickly return to their preexercise state. At an even more extreme level, muscles that are disused for any number of reasons undergo **atrophy**, which can dramatically reduce their functional mass.

Muscles can become disused for a variety of reasons, many of which involve some sort of pathology. Classic examples include paralysis after disruption of the motor nerve innervating the muscle (**lower motor neuron lesion**) or damage to the spinal cord or brain (**upper motor neuron lesion**) that results in paralysis even though the motor nerve fibers leading to the muscle are anatomically intact. Simple extended bed rest also leads to muscular atrophy, often referred to as **disuse atrophy**. The advent of space flight brought into focus an entirely new form of muscular atrophy resulting from extended periods of weightlessness. Animal research into this phenomenon has produced experimental models of weightlessness, a good example being a hindlimb suspension model of rodents in which the animal is suspended by its tail in order to remove any weight-bearing from its hind limbs.

Within recent decades, the phenomenon of **sarcopenia** has attracted considerable attention. Sarcopenia—etymologically "a deficiency of flesh"—refers generically to a reduction of muscle mass. Originally used to describe the loss of functional muscle mass that occurs during aging, sarcopenia is now recognized as a problem accompanying a variety of disease conditions, such as many types of tumors or chronic heart and kidney disease (Fig. 5.1).

Given the large variety of conditions leading to muscle atrophy, it is not surprising that there are some differences in how muscles respond. Nevertheless, what they all have in common is that the balance between protein synthesis and degradation is tipped in favor of the latter. Atrophy involves not only the muscle fibers, but also the connective tissue components of the muscle. As a general rule, the ratio of connective tissue to muscle fiber mass increases considerably during muscle atrophy.

Conditions leading to muscle atrophy

Denervation

Denervation results in immediate paralysis of muscle, and unless only a single muscle is denervated the muscle is normally not subject to passive stretching or shortening.

Muscle Biology
ISBN 978-0-12-820278-4, https://doi.org/10.1016/B978-0-12-820278-4.00006-2

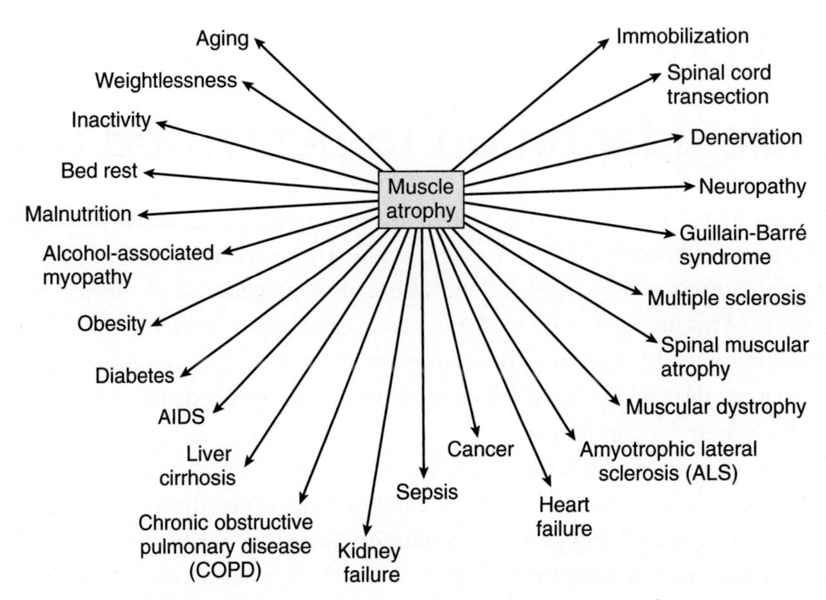

Figure 5.1 Conditions associated with muscle atrophy.

Why profound atrophy follows denervation had been subject to much speculation and research in former decades. Two competing theories dominated the discussions. One posited that transection of the motor nerve cut off the supply of a neurotrophic factor that maintained the muscle fibers. The other focused on the electrical nature of the neural impulse. Experiments showing that the mass and contractile properties of a denervated muscle could be fully maintained simply by electrically stimulating the muscle finally confirmed the validity of the latter hypothesis. Because denervation produces such profound atrophy, this model has proven to be very valuable in understanding mechanisms underlying muscle atrophy and will be treated here in some detail.

Morphological changes. In response to denervation, a muscle undergoes a rapid loss in mass, but an even greater loss in contractile force (Fig. 5.2). Individual muscle fibers lose most of their cytoplasmic volume within months after denervation (Fig. 5.3). The number of myonuclei per muscle fiber declines greatly (one nucleus per day in rats), but in rodents, at least, the number of satellite cells per muscle fiber actually doubles by 2 months after denervation. The reduction in the number of myonuclei per muscle fiber is generally attributed to apoptotic death (Fig. 5.4), but exactly why they die is not yet fully understood. According to the myonuclear domain hypothesis (see p. 63), as the cytoplasmic volume decreases after denervation, there remain more nuclei than are necessary to maintain the existing cytoplasmic volume. Reducing the number of myonuclei would allow the atrophying muscle fiber to remain in an appropriate

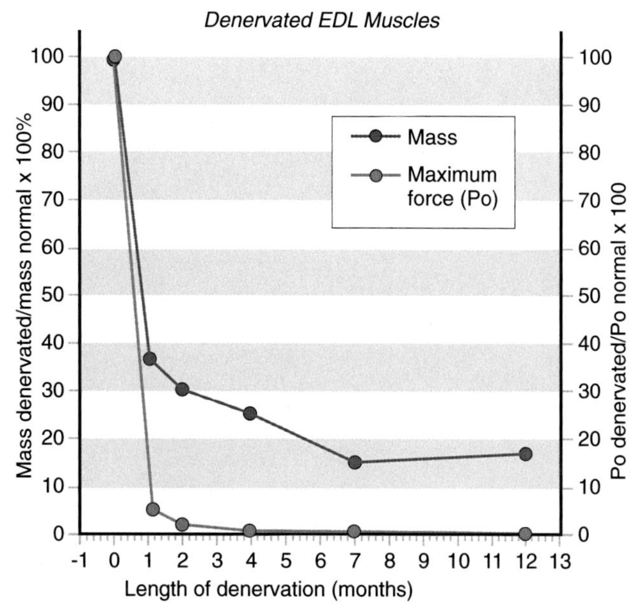

Figure 5.2 Loss of mass and force over time in the denervated rat extensor digitorum longus muscle.

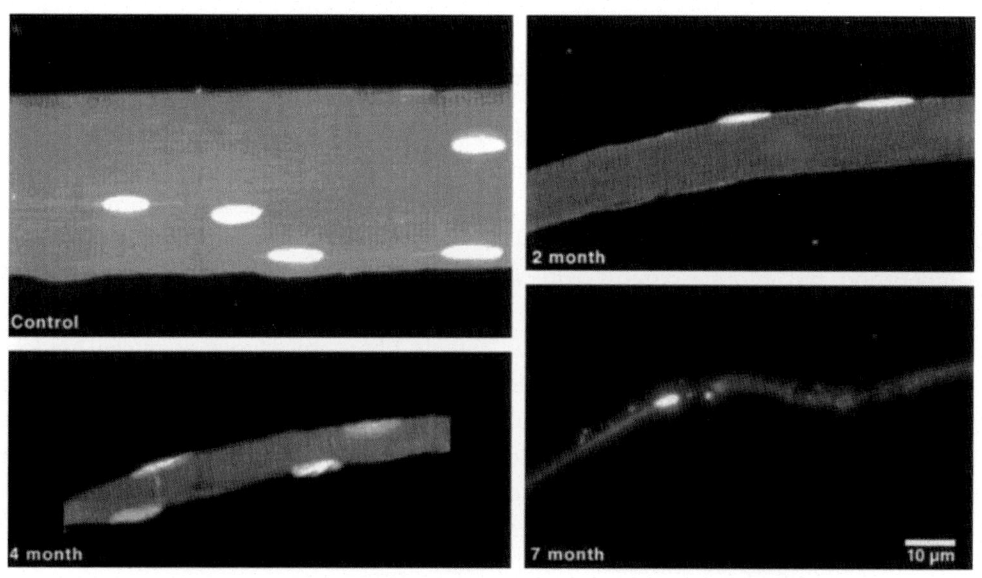

Figure 5.3 Images of denervation atrophy in isolated rat muscle fibers.

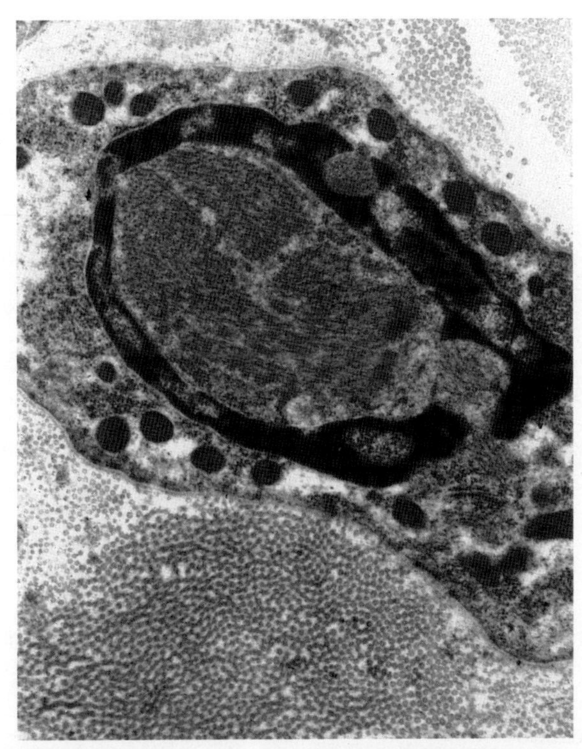

Figure 5.4 Electron micrograph of an apoptotic nucleus in a long-term denervated muscle. Dense masses of cross-sectioned collagen fibers surround the atrophic muscle fiber.

relationship to the diminishing amount of sarcoplasm. By 2 months of denervation, however, the ratio of myonuclear number to cytoplasmic volume is tremendously reduced, suggesting that more variables than the nuclear/cytoplasmic volume ratio are involved. The increase in satellite cell numbers after denervation is also not well understood, but in other systems denervation can result in a temporary increase in cell proliferation. In muscle, levels of certain myogenic factors, for example, MyoD increase after denervation. Long-term denervated muscle fibers are devoid of nuclei for long stretches, interrupted by clumps of nuclei between nonnucleated stretches of cytoplasm (Fig. 5.5). Such clumps of nuclei could be the result of localized proliferation of remaining satellite cells. Although muscle fiber atrophy is dominant in the early months after denervation, many long-term denervated muscle fibers show overt degenerative changes. Surprisingly, the intrafusal fibers of muscle spindles appear to be less affected by denervation than are extrafusal muscle fibers.

Despite its overall atrophy, a denervated muscle remains active at the cellular level. As original muscle fibers undergo atrophy, new muscle fibers form—first (within weeks) between atrophying muscle fibers and their overlying basal laminae and somewhat later (within months) within empty basal laminae of degenerated muscle fibers.

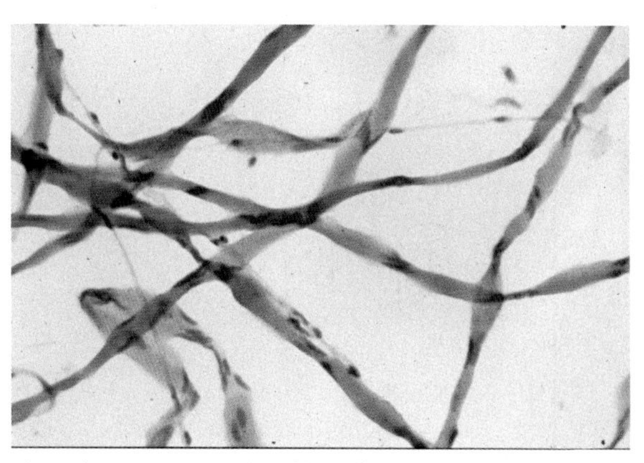

Figure 5.5 Nuclear clumping in teased fibers from a 7-month denervated rat muscle.

Not all types of muscles react the same way to denervation. In rodents, the rate of atrophy is greater in slow than in fast muscles. Actual muscle fiber loss is also greater in slow than in fast muscles. Individual muscle fiber types undergo atrophy at different rates that are not always reflected in the rate of gross atrophy of the muscle. For example, in a fast rat muscle (extensor digitorum longus), Type II muscle fibers, which normally have a larger cross-sectional area than Type I muscle fibers, become atrophic at a faster rate than do their Type I counterparts (Fig. 5.6). For reasons yet unexplained, rates of atrophy of specific muscle fiber types vary according to species, type of muscle and age.

After denervation, one sees the systematic dismantling of the ultrastructural architecture of the muscle fibers. These changes affect both major components of the contractile machinery—the highly organized aggregates of contractile proteins themselves and the structural basis for excitation-contraction coupling (the sarcoplasmic reticulum and the t-tubules). Sarcomeric organization becomes disrupted, and myofibrils are lost. Although the sarcoplasmic reticulum also becomes disorganized, the terminal cisternae, which abut the t-tubules, actually hypertrophy. Mitochondria are also gradually lost. Some of the disruptions seen within denervated muscle fibers are due to the breakdown of desmin and the other sarcoplasmic proteins that maintain the highly organized architectural structure of a normal muscle fiber. (See p. 110 for the molecular events underlying this morphological pattern of atrophy.)

Physiological changes. Corresponding to the disruption of sarcomeric organization, denervated muscles lose contractile force to an even greater extent than overall mass after denervation (see Fig. 5.2). During the same time when contractile force is diminishing, speeds of contraction also commonly change. Although the changes from normal in contraction times vary greatly among species and specific muscles, a common reaction for a fast muscle is for its speed of contraction to become slower. Whereas this may, in part, be due to the more rapid atrophy of Type II muscle fibers (see Fig. 5.6), many other changes within the denervated muscle fibers likely contribute to the change.

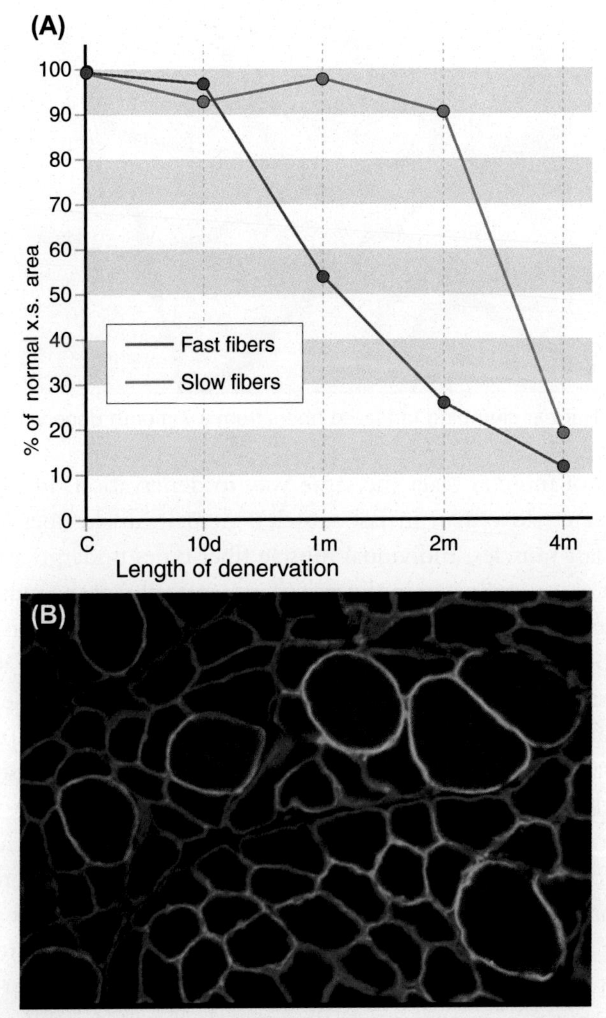

Figure 5.6 (A) Time course of atrophy of fast and slow muscle fibers in denervated rat muscle. (B) Photomicrograph showing greater atrophy of fast versus slow muscle fibers in 2-month denervated rat muscle. Slow muscle fibers are stained brown. The green stain (for laminin) outlines fast muscle fibers.

In contrast, contraction speed tends to increase in muscles that are composed predominantly of slow (Type I) muscle fibers. The increased speed of contraction seen in denervated slow muscles is likely a reflection of a greater dependence upon innervation for the production of slow myosin than fast myosins.

Changes in the composition of ion channels result in a significant reduction in the resting membrane potential of a denervated muscle fiber. One of the most characteristic features of a newly denervated muscle is **fibrillation**—spontaneous uncoordinated contractions that can be seen with the naked eye. Many factors may contribute to fibrillation, but changes in properties of the sarcolemma are likely major contributors to this phenomenon.

One striking change in the denervated sarcolemma is the spread of acetylcholine receptors (an embryonic type) beyond the neuromuscular junction and throughout the surface of the muscle fiber. This represents an almost mirror-image of events that occur during the development of a muscle fiber. In the embryo, early muscle fibers are covered with acetylcholine receptors throughout their length until they become innervated by motor nerve fibers. Upon innervation, acetylcholine receptors become restricted to the forming neuromuscular junction. Accompanying the spread of acetylcholine receptors following denervation is a significant reduction in the number and size of the post-junctional folds in the motor endplate region.

Changes in the microcirculation. Denervation does not affect only the muscle fibers. Blood vessels, both capillaries and the small arteries that feed into them, undergo profound degenerative changes after denervation, but many of these changes may reflect the functional inactivity of a denervated muscle, rather than a dependence of the blood vessels upon innervation. In a study on rats, the capillary supply around denervated muscle fibers was found to decrease by over 4% per week for the first 4 months post-nerve transection (Fig. 5.7). The degeneration and loss of capillaries precedes that of muscle fibers, suggesting that the atrophy and ultimate degeneration of denervated muscle fibers may be due to insufficient oxygenation and metabolic support, as well as the lack of neural stimulation. Accompanying the loss of microcirculation is a great increase in the amount of collagen fibers within the interstitial connective tissue (Fig. 5.8). The surrounding of remaining capillaries and arterioles by dense sheaths of collagen accentuates further the reduction in circulatory support given to the denervated muscle fibers.

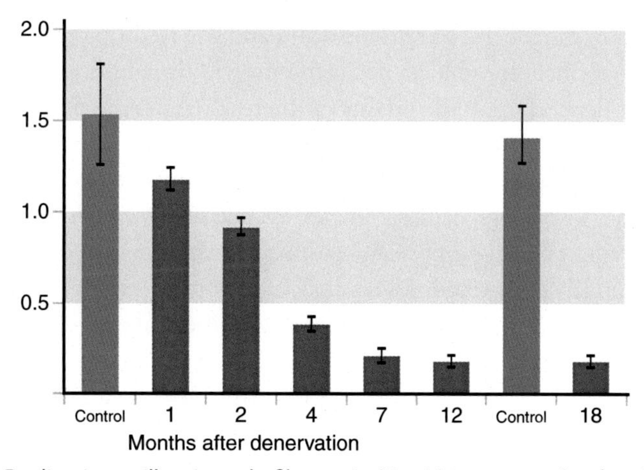

Figure 5.7 Decline in capillary/muscle fiber ratio (Y-axis) in rat muscle after denervation.

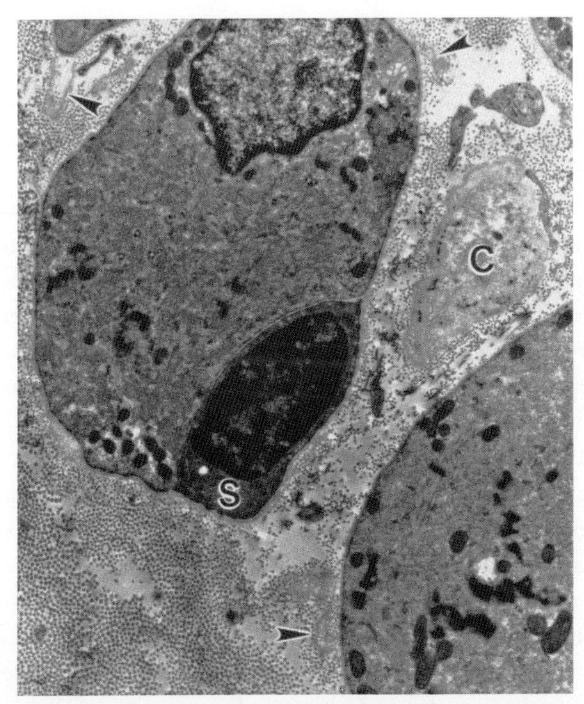

Figure 5.8 Electron micrograph of an 18-month denervated rat extensor digitorum longus muscle, showing a degenerating capillary (C) and basal laminae remaining from completely degenerated capillaries (*arrowheads*). *S*, satellite cell adjacent to a highly atrophic muscle fiber. The gray stippling represents cross-sectioned collagen fibers.

Effects of long-term denervation. Many of the changes seen after denervation of a muscle become greatly accentuated when the period of denervation becomes greater than 6–8 months. These changes are of practical importance because they can have a significant effect upon the success of reinnervation and the restoration of muscle function if regenerating nerve fibers are able to reach the muscle. In addition, new electrical stimulation protocols depend upon the quality of the residual muscle tissue for the restoration of the functional capacity of long-term denervated muscle.

After several months of denervation, the original muscle fibers become extremely atrophic and have lost many of their nuclei. Many of the original muscle fibers have degenerated, leaving behind empty basal laminae or thin newly regenerating muscle fibers located within the persisting basal laminae of the original muscle fibers. In rats, after 4 or 5 months of denervation, the formation of new muscle fibers greatly diminishes while atrophic original muscle fibers continue to degenerate. Surprisingly, muscle spindles showing remarkably little atrophy persist in long-term denervated muscle.

The deposition of large amounts of interstitial connective tissue in the form of dense mats of collagen fibers and deposits of fat cells (adipocytes) is a hallmark of chronically denervated muscle (Fig. 5.9). This is a result of the proliferation of fibroblasts after a

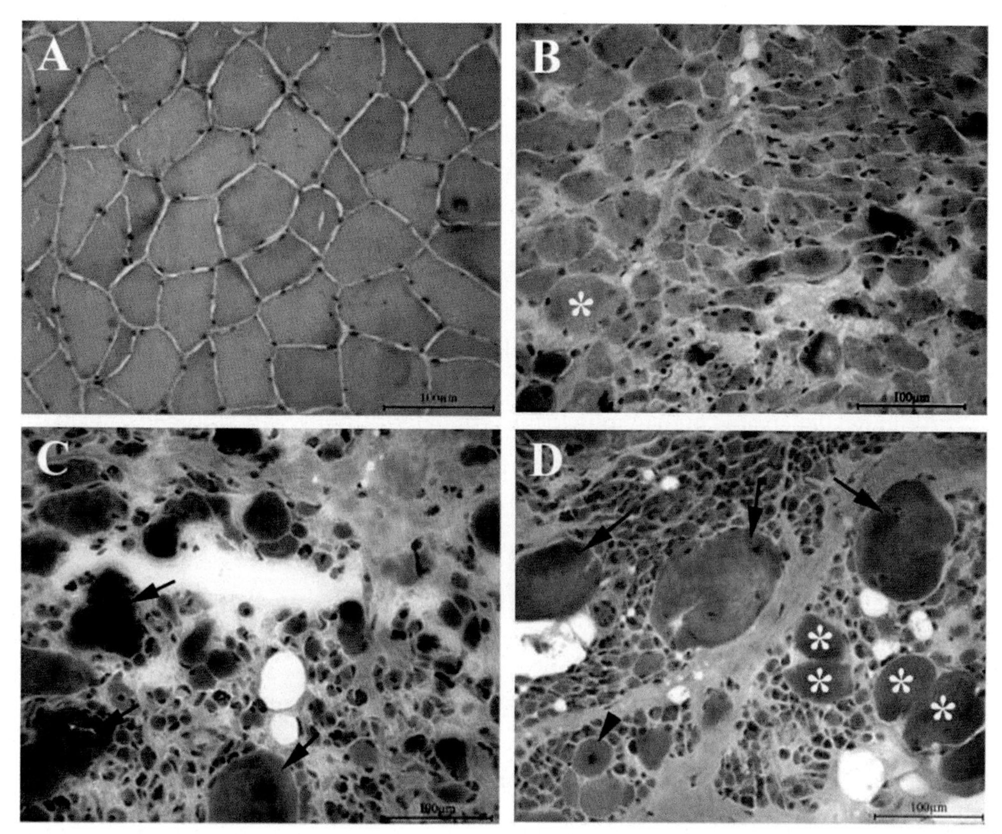

Figure 5.9 Histological sections of long-term denervated human muscles (B—D), showing increasing amounts of muscle fiber atrophy and increasing amounts of interstitial connective tissue and fat. In humans, some large muscle fibers (*arrows and asterisks*) are prominent, but their origins remain uncertain. A - control muscle *Arrowhead*, central nucleated muscle fiber. LML refers to time after complete denervation. *(From Biral et al. (2008), with permission.)*

muscle has been denervated. Whether the increase in the proportion of fat cells is due to an actual increase in their number or their concentration as the muscle atrophies remains to be determined.

Concurrent with the atrophy and degeneration of muscle fibers, the microcirculation undergoes major changes. Within a year after denervation, the capillary/muscle fiber ratio drops 10-fold (see Fig. 5.7). Equally striking, the remaining capillaries become increasingly separated from the atrophic muscle fibers by dense deposits of collagen fibers. As a result, the diffusion distance between the capillaries and atrophic muscle fibers is considerably increased, making the exchange of oxygen and metabolites between the muscle fibers and the microcirculation more difficult. In any part of the body, the microcirculation is highly adaptive to environmental changes, and in long-term denervated muscle there is evidence of considerable remodeling of the microvascular bed.

Significant changes also take place within the intramuscular branches of the motor nerves. Schwann cells, which play a vital role in the regeneration of nerve fibers especially by producing growth factors, persist in muscles denervated as long as 2 years. The major problem in long-term denervation is the obliteration of the channels previously occupied by motor nerve fibers by collagen deposits, effectively cutting these off as conduits between regenerating nerve fibers and atrophic muscle fibers.

Interestingly, clinical studies on long-term denervated human muscle have shown a somewhat different response to long-term denervation from that seen in rodents, the subjects of most research efforts. Understanding these differences is important because of the use of electrical stimulation protocols in attempts to restore some degree of function to these muscles.

Although denervated human muscle fibers undergo atrophy, as do those of rodents, the rate of atrophy is considerably less than that in rats, and the rate of decline in force production is less. Even after a year of denervation, the degree of muscle fiber atrophy in a denervated human muscle is much less than that seen in rodents after only a few months, and the level of fibrosis is also much less (see Fig. 5.9). After several years of denervation, muscle fiber atrophy is much more extreme and the amount of interstitial connective tissue and fatty deposits is correspondingly increased, but interspersed among the atrophic fibers are much larger muscle fibers, some of which appear to be regenerated. As is the case with rats, myonuclear clumping is also seen in long-term denervated human muscles.

Understanding the dynamics of long-term human denervated muscle is important because it can guide attempts to restore function to paralyzed human muscles. Because of fibrosis of nerve channels, attempts to restore innervation of denervated muscle are rarely successful after a year of denervation. However, improvements in electrical stimulation protocols have now made it possible for atrophic muscles to recover a significant degree of function. This technique is surprisingly effective when applied up to a year after denervation, but as might be expected from the pattern of muscle fiber atrophy, the level of success declines as the period of denervation before treatment increases.

Conditions leading to atrophy of innervated muscles

Other than denervation, almost all other conditions leading to muscle atrophy (other than some disease-induced) take place in the presence of nerves. A common denominator in these conditions is disuse. Because the levels and types of disuse vary considerably, the extent of atrophy also varies accordingly. Nevertheless, as will be seen in the last section of this chapter, the fundamental mechanisms underlying the atrophic response of muscle involve some common metabolic pathways.

Spinal cord injury. A spinal cord can become injured through either trauma or disease. In either case, the result can be either partial or complete interruption of the **upper motor neuron** in the central motor pathway. Regardless of the extent of upper motor neuron destruction, the central feature of this condition is the preservation of the **lower**

motor neurons, which directly supply the muscle fibers. Unfortunately, the responses to spinal cord transection often differ between muscles of laboratory animals and humans, so it is not always possible to compare directly clinical data with that obtained from laboratory studies.

A significant feature of upper motor neuron lesions is spasticity of the affected muscles. In laboratory animals, spasticity follows a short period of flaccidity, whereas in humans, flaccidity may or may not precede a spastic phase. In contrast to the flaccid paralysis due to denervation, the spasticity seen after spinal cord lesions is due to the presence of poorly controlled lower motor neurons that stimulate the muscle fibers. This changes the characteristics of the atrophic process.

In humans, spinal cord damage is followed by a rapid phase of muscle fiber atrophy, lasting a few months. Thereafter the level of atrophy remains remarkably stable at between 40% and 50% of the original muscle mass, a marked contrast to denervation atrophy, which is progressive over time. Accompanying a reduction in muscle fiber size is an increase in the proportion of interstitial connective tissue, including more fat cells. The clumping of nuclei, so characteristic of denervated muscle fibers, is not seen. Even though there is some disruption of the ultrastructure of the muscle fibers, the degree of disruption is far less than that seen in denervated muscles.

As would be expected, these atrophic muscles become weaker, but also stiffer possibly as a result of the increased amount of interstitial connective tissue. In laboratory studies, slow muscles are more greatly affected than fast muscles after a spinal cord lesion. In both cases, the speed of contraction becomes faster, but in contrast to the minor change in a fast muscle, the slow soleus muscle becomes populated almost entirely by fast muscle fibers.

Overall, muscles affected by upper motor neuron lesions are much better candidates for rehabilitation than are denervated muscles, not only because of the lesser degree of atrophy, but because of the lesser degree of disruption of the sarcomeric structure within the atrophic muscle fibers. Interventions, such as weight-bearing (treadmill) and electrical stimulation, are more productive in such muscles.

Disuse atrophy. Disuse atrophy is a catchall term that for humans includes the muscular atrophy, resulting from conditions such as immobilization (casting), bed rest, and unilateral lower limb suspension. Most of the data for this category of muscle atrophy have been collected on humans, although immobilization studies based on casting have also been performed on laboratory animals. Unfortunately, the findings from animal studies do not always correspond to those obtained on humans. This discussion will concentrate on human responses to disuse.

Of the common conditions that result in disuse atrophy of muscles, immobilization produces the most profound effects, with about twice the degree of atrophy as that produced by either bed rest or lower limb suspension (Fig. 5.10). Immobilization that eliminates movement across a joint produces the most significant degrees of atrophy.

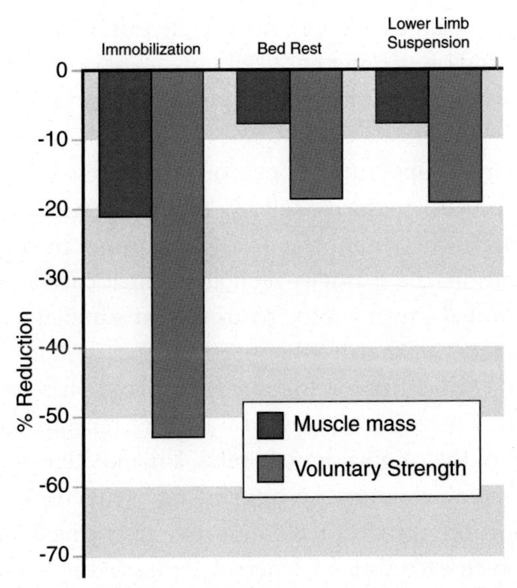

Figure 5.10 Loss in muscle mass and strength following three varieties of disuse. *(Redrawn from Clark (2009).)*

Although muscle mass declines significantly after only a few weeks of immobilization, voluntary strength declines much more than does mass. Not all muscles are equally affected by conditions leading to disuse atrophy. Antigravity muscles (the plantar flexors—soleus and gastrocnemius muscles) are most affected, and muscles of the lower limb experience more atrophy than muscles of the upper limbs or back (Fig. 5.11).

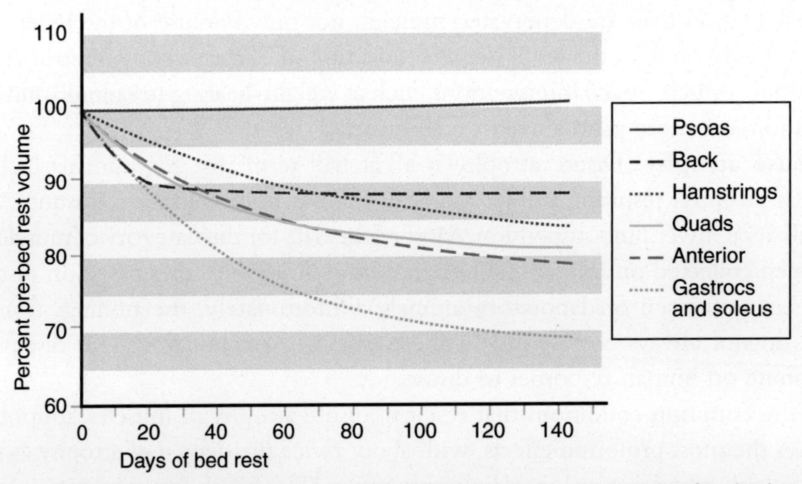

Figure 5.11 Decline in volume of various muscles in the human at periods after bed rest. *(From LeBlanc et al. (1997), with permission.)*

Both the morphology and functional characteristics of muscles suffering from disuse atrophy are based upon more than the muscle fibers alone. MRI studies on atrophic muscles have shown not only a decrease in muscle fiber cross-sectional area, but also a significant increase in intermuscular adipose tissue. The discrepancy between loss of muscle mass and the considerably greater loss of voluntary strength is explained to a considerable extent by a loss of neural mechanisms, in particular a major deficit in centrally induced activation of the motor system. Disuse atrophy involves not only muscle and nerve, but also tendons. Disuse decreases the stiffness of tendons at a faster rate than the decrement in actual muscle function, suggesting that part of the decreases in strength may be a byproduct of the pull of a tendon on its attached muscle.

Disuse affects individual muscle fibers in different ways. After bed rest, both force and power are more greatly affected in slow than in fast muscle fibers (Fig. 5.12). Reductions in the number and length of some thin filaments in disused muscle may partially account for the reduction in strength. Animal studies suggest that disused muscle is more fatigable. As a correlate, atrophic muscles in mice show a 50% decline in mitochondrial density and ATP production. The reduced number of mitochondria is principally the result of autophagy of damaged mitochondria (see p. 114). Those remaining mitochondria are healthier, but their smaller number could contribute to greater fatigability. Data on human muscle are sufficiently variable that extrapolation from animals to humans may not be appropriate.

Weightlessness. Until the first flight of Sputnik, the effect of weightlessness upon skeletal muscle or any other component of the body was not a highly relevant question.

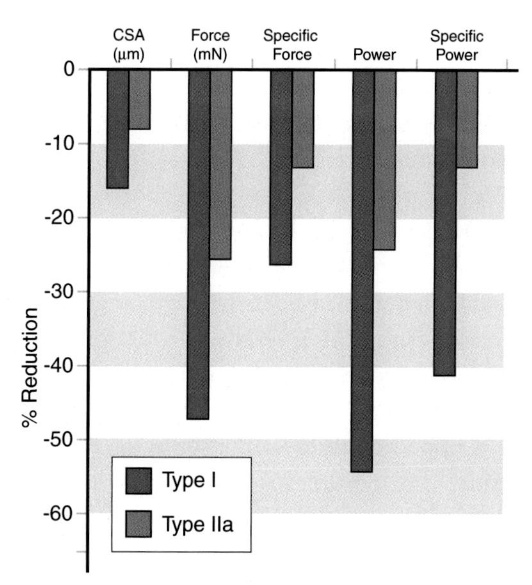

Figure 5.12 Decline in various properties of Types I and II human muscle fibers after 84 days of bed rest. *(Redrawn from Clark (2009).)*

Since then, the increasing amount of space exploration and prospects for long-term space flights to other planets has forced those involved in space programs to pay close attention to the effects of weightlessness upon the human body. For practical reasons, it has been difficult to design clean scientific experiments under actual space conditions. Nevertheless, a general consensus about effects of weightlessness upon various components of the body has emerged. A significant question—beyond the scope of this book—is how to counteract these effects.

Although spaceflight and the accompanying weightlessness affect essentially every bodily system, the effects are especially prominent on the musculoskeletal system, much of which is designed to resist the effects of gravitational pull on earth. When this is missing, the bones, muscles, and ligaments respond quickly to the reduced functional demand. In the case of bone, a decrease in density, with the accompanying loss of calcium, is the most noticeable effect. In the case of muscle, gross muscle atrophy is the most prominent. Tendons lose their stiffness, as well.

Declines in overall muscle mass between 10% and 30% have been reported for space flights between 2 weeks and 6 months, and a steady state is reached by about 6 months. The declines in mass are not similar for all muscle groups. Those muscles most affected by atrophy are the antigravity muscles, for example, the quadriceps muscles in the thigh and the triceps surae (soleus and gastrocnemius muscles) of the lower leg. Muscles of the arms and trunk are less affected. Unfortunately, obtaining "pure" data on gross muscle atrophy, as well as those at a finer level of resolution, is difficult because of the recognition of weightlessness-induced muscle atrophy soon after space flights began. Starting very early in most space exploration programs, astronauts have been advised to carry out resistance exercise programs while in space in order to reduce the extent of atrophy. As a result, it is very difficult to obtain long-term data on prolonged effects of weightlessness on muscle. For this reason, models of bed rest, immobilization, etc. (see above) have served to provide approximations.

Unlike some other components of the body, e.g., bone, weightlessness-induced muscle atrophy is not linear. A relatively rapid rate of atrophy during the first couple of weeks is followed by a slower rate of decline. Fundamental to the process is a reduction in cross-sectional diameter of individual muscle fibers rather than an actual loss of muscle fibers. To complicate interpretation of research results, the rate of muscle fiber atrophy in rodents is close to three times that of humans under conditions of weightlessness. For example, slow Type I muscle fibers are most affected by atrophy in rodents, whereas in humans the balance is tipped more toward atrophy of Type II fibers.

One consistent result of exposure to microgravity is a significant reduction in maximum force and an even greater loss of power (force x time) of muscles and individual muscle fibers in both rodents and humans. The reduction in overall voluntary muscle strength in humans is complicated by neural factors. Even a short exposure to microgravity results in a reduction in neural signals coming to the muscles as determined by electromyographic studies, but in addition there is considerable evidence for a role by central factors that originate in higher centers within the nervous system.

The reduction in maximum tetanic force (P_o) seen in both whole muscles and individual muscle fibers in rodents can be directly related to the reduction in cross-sectional area of muscle fibers. A factor that could contribute to the loss of force and power is the selective loss of thin (actin) filamentous material within the sarcomeres that has been described in both rodents and humans. Another common functional feature is a general speeding up of contraction times, especially in slow muscle fibers, that is correlated with an increase in expression of fast myosin in the atrophic muscle fibers.

Muscles exposed to zero gravity are more fatigable than normal muscle fibers. In addition, there is a shift in overall metabolism from oxidative to glycolytic, as well as a shift from lipids to carbohydrates (glycogen) as a source of energy. Correspondingly, the mitochondrial content of the atrophic muscle fibers is greatly reduced. The increased number of damaged mitochondria is removed through autophagic processes (see p. 113). The oxidative functions of the remaining mitochondria also appear to be reduced.

A significant problem associated with spaceflight is muscle damage upon reentry into a gravitational environment. Many astronauts have complained about persistent muscle soreness after a return from space. The pain resembles that experienced from eccentric muscle contractions. In addition to the reduction in thin filaments (see above), other disturbances in sarcomeric structure predispose space-adapted muscles to damage when gravitational forces are again applied. Such damage is followed by regeneration. In general, muscles exposed to either weightlessness or other disuse conditions recover their original mass and functional characteristics after return of the subject to normal earthbound activities.

Disease. A number of diseases are characterized by muscle atrophy that is based upon more than simply disuse, although it is not always possible to separate the two (see Fig. 5.1). Among these diseases are diabetes, congestive heat failure, hyperthyroidism, uremia, many neoplastic conditions, and infectious diseases. Even though they do not fall strictly into the category of disuse atrophy, they are nevertheless mentioned here because many of the mechanisms underlying disease-based atrophy appear to be similar to those operating in disuse atrophy. Nevertheless, there are also significant differences. Many of the mechanisms underlying disease-based muscle atrophy are responses to systemic factors that are not prominent in muscle atrophy due to disuse. Among these factors are inflammatory cytokines and stress hormones (e.g., cortisol) circulating in the blood. Complicating things even further, many people with chronic diseases are confined to bedrest or have considerably reduced physical activity, so the muscle atrophy seen in these conditions represents a convergence of systemic and local factors.

The metabolic and molecular basis of muscular atrophy

The loss of functional muscle mass during atrophy is the net result of many interacting molecular and metabolic factors. At a most basic level, it reflects a balance between protein synthesis and degradation, with the net result being less protein in the atrophic muscles. Such a net negative balance, however, could be brought about in several ways, for example, a major decrease in protein synthesis without a significant change in the rate of degradation or, conversely, a greatly increased rate of degradation in the face of a normal rate of protein synthesis. Another, and probably most likely, scenario is the combination of reduced synthesis and increased degradation of muscle proteins. What is clear is that during the early phases of unloading muscle loses mass rapidly, but then enters a plateau phase during which the remaining mass is lost only slowly.

Although our understanding of the internal factors leading to muscle atrophy is far from complete, it is apparent that no single atrophy program covers all forms of disuse atrophy. In addition, even under a single condition of disuse, not all muscles react in the same manner. To complicate things even further, there are significant differences between how human and laboratory rodent muscles react to the same unloading stimulus. Nevertheless, a general picture of the molecular underpinnings of muscle atrophy is beginning to emerge.

Regardless of the balance between protein synthesis and degradation, unloading a muscle stimulates a molecular atrophy program that results in the dismantling of many of the sarcomeric proteins. Four main degradative pathways are involved in the breakdown of muscle proteins: **Ca^{++}-dependent proteases** (**calpains**), **caspases**, the **ubiquitin proteasome pathway (UPP),** and the **autophagy** (lysosome) system. Of these, the UPP pathway plays a dominant role in muscle protein breakdown.

UPP pathway

A key to activation of the UPP pathway is the transcription factor family **FoxO** (Fig. 5.13). Under conditions of growth or healthy stability, the activity of FoxO is kept in check through inhibition by Akt1. The inhibition by Akt1 is accomplished through the phosphorylation of FoxO, which sequesters the FoxO molecules in the cytoplasm and renders them ineffective as modifiers of gene expression. Under unloading conditions leading to muscle atrophy, a number of types of input decrease the inhibitory influence of Akt1 upon FoxO by reducing its phosphorylation. In its nonphosphorylated form, FoxO moves into the nucleus, where it activates a family of genes (**atrogenes**) that promote the degradation of sarcomeric proteins.

Understanding the mechanisms underlying muscle atrophy is complicated by systemic factors that can also attenuate the functions of Akt1 or activate FoxO. Examples of such systemic factors are cytokines resulting from inflammatory processes and circulating hormones (e.g., cortisol), which are increased because of pathologic processes

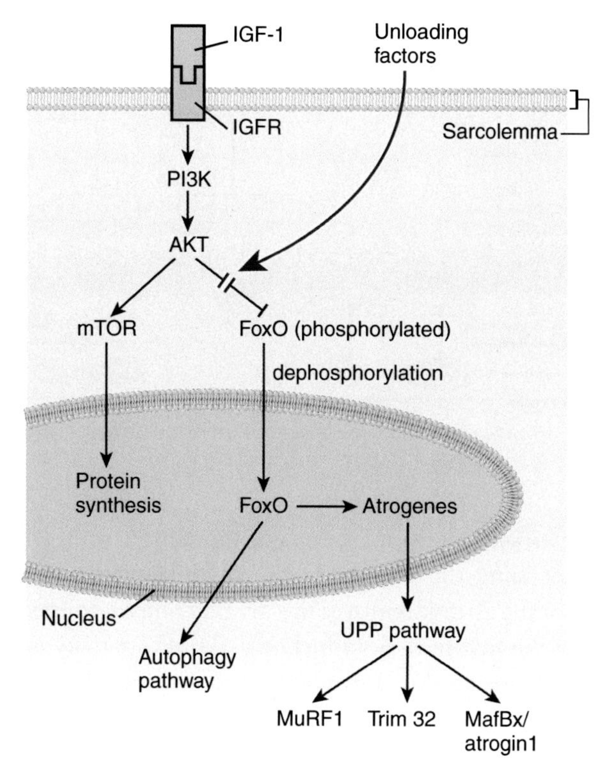

Figure 5.13 Major pathways involved in muscle fiber atrophy. *AKT*, serine/threonine protein kinase 1; *IGF-1*, insulin-like growth factor; *IGFR*, insulin-like growth factor receptor; *TOR*, target of rapamycin; *UPP*, ubiquitin phosphate.

affecting the entire body. These systemic influences on atrogene products will not be covered in this chapter (see Chapter 7). Instead, only factors relating to unloading and disuse are considered here.

The atrogenes and their protein products that are activated by disuse play several different roles in the atrophic process. Most prominent is the degradation of sarcomeric proteins. The UPP pathway and its atrogenes are responsible for ~75% of the protein breakdown within atrophying muscle fibers. Another seemingly paradoxical function of atrogenes is their influence on various aspects of myogenesis—both stimulatory and inhibitory. A third general function of atrogenes is the modulation of **autophagy**—the removal of damaged cytoplasmic organelles and proteins by lysosomal activity.

Three atrogene products that are released through the action of FoxO play a major role in the disassembly of sarcomere structure during atrophy. These are **MuRF1, MafBx/ atrogin1**, and **Trim32**. Each of these plays a different, but important role (Fig. 5.14). The activity of MuRF1 focuses on the thick filaments and targets myosin heavy chains, myosin light chains myosin binding protein C and possibly troponin on the thin filaments.

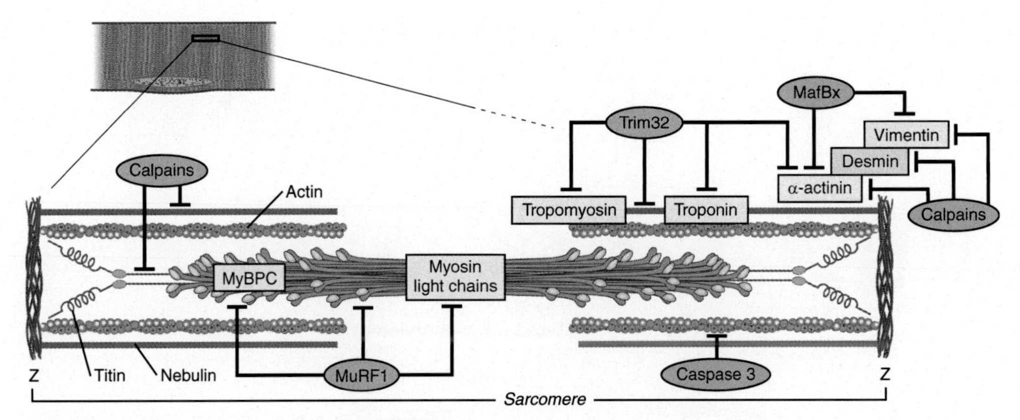

Figure 5.14 Various enzymes (ovals) and their targets in the breakdown of sarcomeric proteins during muscle atrophy.

In contrast, MafBx/atrogin1 degrades the major proteins, desmin and vimentin, that are responsible for maintaining the register of the myofilaments and the structure of the Z-disk. Dissolution of the Z-disk removes the intersarcomeric connections and effectively makes that region of the myofibril nonfunctional. Trim32 activity is directed toward the thin filaments, where it breaks down troponin, tropomyosin, α-actinin, and possibly also actin filaments. These atrogene proteins are all members of the E3 group of ubiquitin enzymes (see p. 70 and Fig. 3.9), and they transfer the partially degraded muscle proteins to proteasomes, where the proteins are further broken down into simpler molecules.

Atrogens also play a role in myogenesis. Atrogin1 targets the myogenic regulatory factors, MyoD and myogenin (See Fig. 2.11), for destruction by proteasomes. Although MyoD and myogenin do not act upon mature muscle fibers, their degradation via the UPP system would interfere with any new muscle fiber formation in disused muscle. Paradoxically, MyoD and myogenin are both induced in denervated muscle, and myogenin itself activates promoters for the MuRF1 and atrogin1 genes.

Calpains and caspases

Both **calpains** (Ca^{++}-activated proteases) and **caspases** (which do not require Ca^{++} for activation) play an important role in both sarcomeric remodeling in normal muscle and in the atrophy of disused muscle. They show increased activity in denervated muscle. These families of enzymes play an important role in the initial disassembly of myofibrils in both disused and remodeling muscle fibers. Instead of attacking the proteins of the thick and thin filaments, calpains, along with MafBx/atrogin1, focus on the degradation of proteins of the Z-disk and other supporting sarcomeric proteins, such as desmin, vinculin, talin, nebulin, and titin (see Fig. 5.14). Removal of the Z-disk exposes the thin filaments to the action of caspase 3, which along with Trim32 begins the initial breakdown of actin

BOX 5.1 The role of calpains and caspases in meat tenderization

Some of the enzymes involved in the muscle atrophy program maintain their activity after death, and in the meat industry, they play important roles in the process of meat tenderization. Especially important is the calpain system, which includes three calpain isoforms, μ-calpain,[1] m-calpain, and calpain 3, along with a specific endogenous calpain inhibitor, **calpastatin.** Of these isoforms, μ-calpain and m-calpain are of greatest importance in the meat tenderization process. μ-Calpain is the most active during the early days following slaughter, whereas the levels of Ca^{++} are not optimal for the activation of m-calpain postslaughter. Increasing tenderness of meat is due principally to the enzymatic breakdown of the sarcomeric proteins (e.g., desmin, titin, nebulin, troponin-T) that hold the myofilaments in place, rather than the myosins and actin of the myofilaments themselves. The postmortem activities of the calpains are themselves inhibited by the activity of calpastatin, and meat containing high levels of calpastatin is tough and of poor quality.

The role of caspases in postmortem meat tenderization is less clear, but research on callipyge ("beautiful hind-end") lambs, which are noted to have tough meat, has shed some light on their importance. Calpastatin levels are high in callipyge lambs as compared to normal lambs. Normal lambs have higher activities of postmortem caspase activity than do callipyge lambs, and there is evidence that this caspase activity degrades calpastatin, thus allowing greater calpain activity and consequently, more tender meat. In callipyge lambs, calpastatin levels are so high that the endogenous caspase activity is not sufficient to reduce the activity of this calpain inhibitor to a commercially sufficient level in this strain of sheep.

[1] The μ- and m-refer to micro- and millimolar amounts of Ca^{++} required to activate these calpains.

filaments before the actin fragments are transferred to the UPP system for further breakdown by the proteasomes. Disruption of the actin filaments and the large myofibril-organizing proteins, titin and nebulin, then opens the myosin proteins of the thick filaments to the degradative actions of MuRF1. Along with their role in the breakdown of sarcomeric proteins, caspases are vital components of the apoptotic pathway that leads to the removal of myonuclei during muscle fiber atrophy. In addition to their roles in muscle atrophy, both calpains and caspases have been identified as being important in the tenderization of meat (see Box 5.1).

Autophagy

Autophagy works at a higher level of organization than the other degradative pathways that operate during muscle atrophy. The essence of autophagy is the envelopment of damaged mitochondria or protein aggregates by a double membrane to form a **phagophore** (Fig. 5.15). Critical to the initiation of the envelopment phase is the regulatory protein **Beclin 1**. The phagophore then merges with a lysosome, exposing the materials enclosed within the phagophore to the wide variety of proteolytic enzymes contained within the lysosomes. The degradation products are either eliminated from the cell or

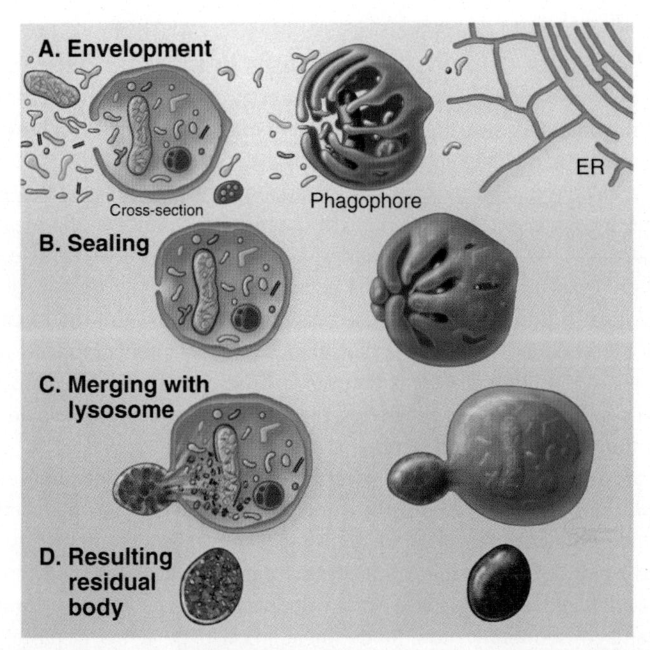

Figure 5.15 Major stages in autophagy. *(From Pollard et al. (2017), with permission.)*

are reutilized for the synthesis of other cytoplasmic macromolecules. Autophagy is an important component of the muscle wasting seen in starvation (Box 5.2).

Noncoding RNAs

To further complicate an already complex picture of degradative pathways leading to muscle atrophy, recent research has shown that **noncoding RNAs** constitute an additional layer of regulatory molecules that influence these pathways. Noncoding RNAs (or micro-RNAs [**miRNAs**]), of which there are many types and hundreds of specific examples, are small RNA molecules (only tens of nucleotides) that regulate gene expression principally at the mRNA (translational) level. It is estimated that 30% of all human genes are regulated by miRNAs. Almost like synaptic connections in the brain, a single miRNA may regulate many target mRNAs, and a single mRNA molecule may be influenced by multiple miRNAs.

Each form of atrophy of skeletal muscle appears to be influenced by different combinations of miRNAs, and different types of muscles are affected by different combinations of miRNAs. Under conditions leading to muscle atrophy, miRNA regulation can depress synthetic pathways or participate in the activation of degradative pathways. Not only atrophy, but myogenesis, regeneration, and hypertrophy are also influenced by miRNAs. Table 5.1 outlines some of the miRNAs that either promote or retard muscle atrophy.

BOX 5.2 Muscle atrophy during starvation

Starvation results in a form of muscle atrophy brought about by neither disuse nor disease. Muscle is greatly affected by starvation because it contains the largest stores of potential internal nutrients of any tissue in the body. The overall response of the body to starvation is to break down both proteins and lipids so that the breakdown products (amino acids and fatty acids) can be reutilized in order to synthesize other molecules critical to overall survival.

Even after 1 day of starvation, at least in laboratory animals, initial adaptations to nutrient deprivation have already begun. Not all components of muscle are equally affected. Fast Type II muscle fibers are more greatly affected than slow Type I fibers. This might be part of an evolutionary response that favors the preservation of postural functions over those that produce voluntary force. The rate of muscle breakdown is greatest during the earliest phases of starvation and is subsequently reduced as the overall mass of the musculature is reduced.

Both the UPP (ubiquitin protease pathway) and autophagy pathways are heavily involved in the wasting of muscle during conditions of starvation, but most of the protein degradation in skeletal muscle occurs through the UPP. Two prominent factors come into play in the stimulation of the UPP, and both affect the IGF-1/Akt/mTOR pathway (see Fig. 3.8). One is a great reduction in insulin/IGF-1 binding to the surface receptors on the muscle fiber, which reduces the strength of this pathway. The other factor is the increase in blood levels of glucocorticoid stress hormones. Both of these factors work to shift the balance of the IGF-1/Akt/mTOR pathway from promoting protein synthetic activities to removing the inhibition of FoxO-3 so that it can translocate to the nucleus and stimulate the expression of genes that control the atrogene system. Then the atrogene proteins, such as MuRF and atrogin1, begin to degrade muscle proteins for internal recycling.

Nutrient deprivation very rapidly stimulates the activators of the autophagy system, which accounts for something less than 25% of the protein breakdown that occurs during starvation. Playing a major role in stimulation of the autophagy pathway is a reduction in the inhibitory influence of Akt upon FoxO-3, as also occurs during the induction of the UPP pathway. FoxO-3 then stimulates the expression of genes that activate the autophagy system.

Table 5.1 Some miRNAs that play a role in muscle atrophy.

Atrophy-promoting miRNAs	Atrophy-retarding miRNAs
General muscle atrophy—miR-206	Spinal muscular atrophy—miR-196a
Denervation atrophy—miR-21, miR-206	Denervation atrophy—miR-351
Many types of atrophy—miR-29b	Kidney disease atrophy—miR-27a
Burn-induced atrophy—miR-628	ALS-related atrophy—miR-206
Dexamethasone atrophy—miR-1	Dexamethasone atrophy—miR-223a

References

Bell RAV, Al-Khalaf M, Megeney LA. The beneficial role of proteolysis in skeletal muscle growth and stress adaptation. Skel Muscle 2016;6:16.

Bilodeau PA, Coyne ES, Wing SS. The ubiquitin proteasome system in atrophying skeletal muscle: roles and regulation. Am J Physiol Cell Physiol 2016;311:C392–403.

Biral D, Kern H, Adami N, Boncompagni S, Protasi F, Carraro U. Atrophy-resistant fibers in permanent peripheral denervation of human skeletal muscle. Neurol Res 2008;30:137–44.

Carlson BM, Borisov AB, Dedkov EI, Dow D, Kostrominova TY. The biology and restorative capacity of long-term denervated skeletal muscle. Basic Appl Myol 2002;12:249–56.

Carlson BM. The biology of long-term denervated muscle. Eur J Trans Myol 2014;24:5–11.

Clark B. *In vivo* alterations in skeletal muscle form and function after disuse trophy. Med Sci Sports Exerc 2009;41:1869–75.

Finn PF, Dice JF. Proteolytic and lipolytic responses to starvation. Nutrition 2006;22:830–44.

Fioletta VC, White LJ, Larsen AE, Leger B, Russell AP. The role and regulation of MAFbx/atrogin-1 and MuRF1 in skeletal muscle atrophy. Pflueg Arch Eur J Physiol 2011;461:325–35.

Fitts RH, Riley DR, Widrick JJ. Functional and structural adaptations of skeletal muscle to microgravity. J Exp Biol 2001;204:3201–8.

Foran JRH, Steinman S, Barash I, Chambers HG, Lieber RL. Structural and mechanical alterations in spastic skeletal muscle. Dev Med Child Neurol 2005;47:713–7.

Gao Y, Arfat Y, Wang H, Goswami N. Muscle atrophy induced by mechanical unloading: mechanisms and potential countermeasures. Front Physiol 2018;9:1–17.

Hoppler H. Molecular networks in skeletal muscle plasticity. J Exp Biol 2016;219:205–13.

Kemp CM, Sensky PL, Bardsley RG, Buttery PJ, Parr T. Tenderness - an enzymatic view. Meat Sci 2010;84: 248–56.

Kern H, Moedlin M, Mayr W, Vindigni V, Zampieri S, Boncompagni S, Protasi F, Carraro U. Spinal Cord 2008;46:293–304.

Midrio M. The denervated muscle: facts and hypotheses. A historical review. Eur J Appl Physiol 2006;98: 1–21.

Mirzoev TM, Shenkman BS. Regulation of protein synthesis in inactivated skeletal muscle: signal inputs, protein kinase cascades, and ribosome biogenesis. Biochemist 2018;83:1299–317.

Narici MV, de Boer MD. Disuse of the musculo-skeletal system in space and on earth. Eur J Appl Physiol 2011;111:403–20.

Ohira T, Kawano F, Ohira T, Goto K, Ohira Y. Responses of skeletal muscles to gravitational unloading and/or reloading. J Physiol Sci 2015;65:293–310.

Pandurangan M, Hwang I. The role of calpain in skeletal muscle. Anim Cell Syst 2012;16:431–7.

Rodriguez J, Vernus B, Chelh I, Cassar-Malek I, Gabillard JC, Hadj Sassi A, Seiliez I, Picard B, Bonnieu A. Myostatin and the skeletal muscle atrophy and hypertrophy signaling pathways. Cell Mol Life Sci 2014; 71:4361–71.

Schiaffino S, Dyar KA, Cicliot S, Blaauw B, Sandri M. Mechanisms regulating skeletal muscle growth and atrophy. FEBS J 2013;280:4294–314.

Siu PM. Muscle apoptotic response to denervation, disuse, and aging. Med Sci Sports Exerc 2009;41: 1876–86.

Xiao J (ed.). Muscle Atrophy. Adv Exp Med Biol 2018;1088:1–624.

CHAPTER 6

Muscle injury and regeneration

During the course of a lifetime, muscle may become injured. Both the type and severity of the injury can vary tremendously, and the response of the injured muscle varies accordingly. In some cases, the injured muscle responds by scarring; in others, the response can be complete regeneration. An example of the latter has already been introduced, namely, the breakdown and subsequent regeneration of muscle fibers injured by eccentric muscle contractions (see p. 79). The response of muscle to severe traumatic injury is much more complex, and to a great extent it is related to the extent of damage to the blood vessels that supply the damaged muscle fibers. Although not affecting the initial stages of regeneration, the type and extent of damage to the motor nerve fibers that supply the damaged muscle fibers can affect the level of functional return of the regenerating muscle fibers.

Regardless of the nature of the initial injury, the regenerative process of mammalian muscle fibers follows a similar pattern, starting with the activation of satellite cells, their proliferation, and subsequent fusion to form regenerating myotubes, which then undergo maturation to form mature regenerated muscle fibers. All of this typically occurs within the persisting basal laminae of the original muscle fibers.

Although seemingly straightforward now, our contemporary understanding of the cellular nature of muscle fiber regeneration is the result of many decades of research. Even though muscle fiber regeneration was described in the German medical literature during the late 1800s, the ability of damaged skeletal muscle fibers to regenerate was forcefully denied in much of the English literature during the first half of the 20th century. This was in large part due to the idea that the nuclei of a mature muscle fiber—a highly differentiated syncytial tissue—are incapable of mitotic division. In addition, pathologists examining previously traumatized human muscle tissue found that the areas of damage had been filled in by dense collagenous scar tissue.

In the meantime, significant muscle regeneration was observed by scientists who were studying the regeneration of amputated salamander limbs. After amputation of a limb (or tail), the tissues near the amputation surface undergo a profound reorganization to form a mass of cells (the **regeneration blastema**) that is the developmental equivalent of the embryonic limb bud (Fig. 6.1A and B). The blastema develops into a new replacement for the part of the limb that had been amputated through a process of **epimorphic regeneration**, and the newly regenerated limb contains a normal skeleton and musculature (Fig. 6.2). Where did the new muscle come from? The dominant hypothesis was that the transected ends of the muscle fibers undergo a fragmentation response

Muscle Biology
ISBN 978-0-12-820278-4, https://doi.org/10.1016/B978-0-12-820278-4.00011-6

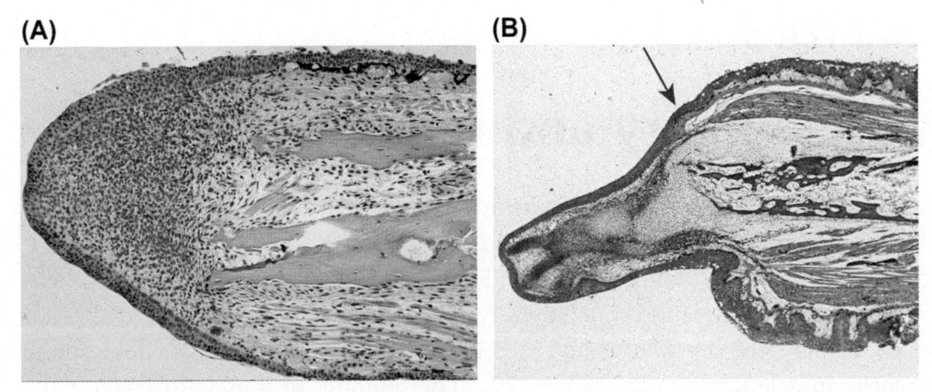

Figure 6.1 (A) Regenerating limb of a newt in the blastema stage. The mass of homogeneous purple cells at the left of the figure is the regeneration blastema. (B) Regenerating limb of an axolotl, showing early development of the main skeletal structures of the limb, including primordia of three digits. The dark cellular masses alongside the skeletal elements are cellular precursors of the musculature. The arrow indicates the plane of amputation.

Figure 6.2 A mature hand regenerate in the axolotl. All skeletal elements of the hand (in the cartilaginous stage) are normal. Regenerated muscles can be seen between the digits.

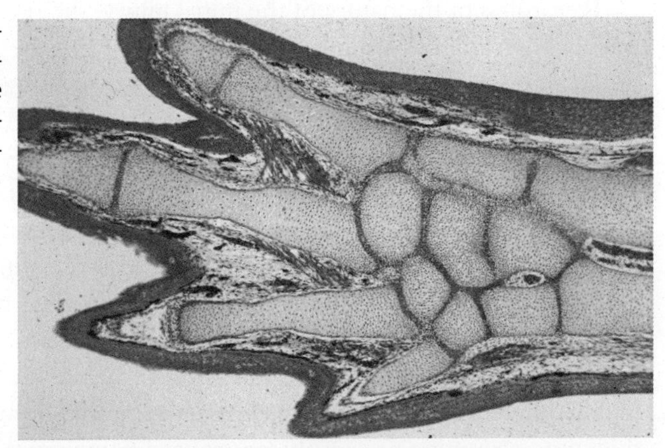

(**dedifferentiation**), and the nucleated fragments became surrounded by a plasma membrane to form the equivalent of embryonic muscle precursor cells (Fig. 6.3). These cells become incorporated into the blastema, and under the influence of the developmental controls inherent in the blastema, they develop into the new musculature of the regenerating limb. In doing so, the regenerating muscles follow a pattern of morphogenesis essentially identical to that which occurs in the embryo.

By the mid-1900s, reports from a number of laboratories demonstrated conclusively that damaged mammalian muscle fibers can also regenerate. The question was how. Two alternatives emerged with the description in 1961 of a nondescript mononuclear cell, called a satellite cell, located between a muscle fiber and its surrounding basal lamina

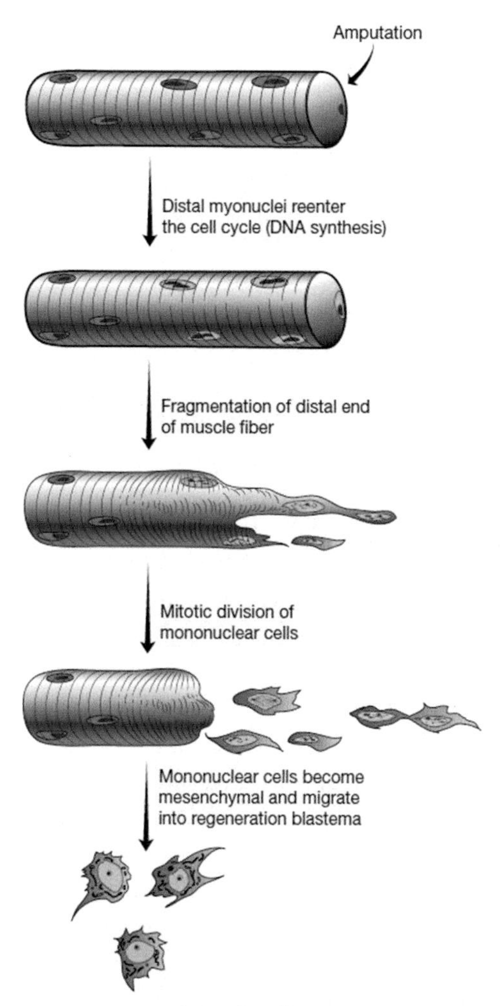

Figure 6.3 Diagrammatic representation of the dedifferentiation of a muscle fiber in an amputated amphibian limb and the conversion of the resulting mononuclear cells into blastemal cells. *(From Carlson (2007), with permission.)*

(see Fig. 1.2). The satellite cell was suggested to be a possible source of the cells involved in muscle fiber regeneration. Thus, there were two alternatives that could account for the origin of regenerating muscle fibers—satellite cells or the dedifferentiation of the injured muscle fiber. This set up a sometime bitter controversy between proponents of the two alternatives. As is often the case with controversies between two competent scientific sides, neither side proved to be completely correct or completely wrong. Even after 50 years, some aspects of this controversy have not yet been definitively decided. Current evidence shows that satellite cells are the main source of the cells involved in mammalian skeletal muscle regeneration, but in nonmammalian species, some muscles regenerate from satellite-like cells, whereas others use dedifferentiation as the source of myoblastic cells for regeneration.

The epimorphic regeneration of skeletal muscle

The epimorphic regeneration of skeletal muscle is complex because it takes place within the overall context of a regenerating limb. Two main questions have dominated research on the epimorphic regeneration of muscle: (1) What is the origin of the regenerating muscle cells? and (2) What factors determine the shape of the regenerating muscles?

Cellular origins. The source of myogenic cells during epimorphic limb or tail regeneration has proven to be much more complex than was envisioned when this process was first studied. Whether new muscle forms through the dedifferentiation of injured muscle fibers or from satellite cells depends upon not only the species, but in some instances the stages in the life cycle of the animal. In several animals, ranging from *Amphioxus* to electric fish to *Xenopus* tadpoles, satellite cells, or their equivalent have been found to serve as the source of regenerating muscle during the overall epimorphic regeneration of the tail. Surprisingly, during limb regeneration, dedifferentiation of multinucleated muscle fibers has been demonstrated in newts, but only in the adult form. On the other hand, newt larvae and both larvae and postmetamorphic adults of axolotls produce new muscle in regenerating limbs through the recruitment of satellite cells.

Control of shape. The blastema of a regenerating limb has many features in common with an embryonic limb bub. A major difference is that the regeneration blastema is connected to the stump of a mature limb (see Fig. 6.1A), which could exert a major influence upon the blastema. The most critical influence is that of nerves. Regenerating nerve fibers penetrate throughout the blastema. In fact, the blastema is dependent upon innervation. Without an adequate nerve supply, a regeneration blastema does not form. Instead, the amputation surface heals over by a scarring process similar to that seen in an amputated mammalian limb.

As the blastema is taking shape, it acquires the morphogenetic information necessary to guide the shaping of both the skeleton and the musculature of the regenerate. If a blastema is removed from the limb stump and is grafted to some other location in the body, it will still develop into a reasonably normal looking limb, thus demonstrating morphogenetic independence from the limb stump. As the limb is regenerating, the development of both the skeleton and muscles follows a pattern remarkably similar to that seen in the embryonic limb bud. Details of the pattern-forming mechanisms are beyond the scope of this book. As will be seen below, embryonic pattern-forming mechanisms do not reemerge in isolated regenerating mammalian muscles (or amphibian muscles for that matter). Instead, mechanical influences determine the overall shape and internal architecture in mammalian muscle regeneration.

The tissue regeneration of skeletal muscle

Because mammalian limbs cannot regenerate, all known types of muscle regeneration in mammals take place through the tissue mode. Regeneration by the tissue mode is a local healing mode that at the cellular level involves the reawakening of embryonic differentiation patterns to go from satellite stem cells to multinucleated muscle fibers. At the level of morphogenesis of an entire muscle, however, embryonic morphogenetic control mechanisms are not called into play as they are in epimorphic regeneration.

The fundamental steps in mammalian muscle regeneration are (1) some form of damage, whether mechanical, vascular, chemical, or disease-induced; (2) infiltration of the area of damage by inflammatory cells; (3) activation and proliferation of satellite cells; (4) regenerative myogenesis; (5) gross and internal architectural modifications (morphogenesis); and (6) functional maturation.

Types of muscle damage leading to regeneration

Muscles and muscle fibers can become damaged in a remarkably large number of ways. How a muscle is damaged is important because the circumstances of injury often determine the shape of the regenerative response or lack thereof. At the level of the muscle fiber, there are few variations in the nature of regeneration, but at the level of muscle as a tissue or organ, differences can be significant.

Mechanical trauma. Mechanical trauma is a major cause of muscle damage, but trauma can occur in a variety of dimensions. Gross trauma, resulting from accidents or bullet wounds, for example, frequently causes severe muscle damage, often involving a significant loss of tissue. Such forms of trauma are among the hardest in which to mount a functionally effective regenerative response because large masses of muscle tissue are lost. This deprives regenerating muscle cells of a suitable substrate within which to replace the lost tissue. As a result, such wounds are often replaced by extensive scarring, with only a minimal amount of regenerating muscle along the edges. Other gross trauma, such as that resulting from blunt blows to the body, can result in muscle fiber damage, but because no tissue is actually lost through the blow, the basic tissue architecture that would support muscle fiber regeneration is often preserved. Of particular importance in gross mechanical trauma is disruption of the motor nerve supply. If the nerves leading to a severely damaged muscle are avulsed, adequate nerve regeneration must accompany that of the muscle in order for functional regeneration of the muscle to occur. Although the early stages of muscle fiber regeneration are nerve-independent, the regenerating muscle fibers will undergo atrophy in the absence of innervation.

Other forms of trauma involve even less tissue disruption. The damage to muscle fibers resulting from eccentric muscle contractions has already been discussed (see p. 81). In this case, the damage to both muscle fibers and interstitial connective tissue does not disrupt the overall organization of the muscle tissue, and the microvasculature remains largely intact. Such conditions are optimal for a robust regenerative response.

Some muscle diseases, especially muscular dystrophy (see Chapter 7), leave the muscle fibers susceptible to microtrauma. Mutations in dystrophin or certain other structural molecules leave muscle fibers in a weakened state in which the sarcolemma is not capable of withstanding the mechanical forces imposed upon them by many ordinary body movements. As a result, these movements can rupture the membranes covering the muscle fibers, leading to muscle fiber damage.

Vascular damage/ischemia. If muscle fibers are deprived of an active blood circulation for more than a couple of hours, they die. Satellite cells are somewhat more resistant to ischemia than muscle fibers, but in the absence of a blood supply, they also die. About 4 hours of ischemia causes irreversible damage to muscle fibers and satellite cells (**ischemic necrosis**) in both humans and laboratory animals (Fig. 6.4A). Under conditions of ischemic necrosis, the dead muscle fibers remain essentially unchanged until blood vessels begin to grow into the ischemic area. This sets up a gradient of muscle fiber degeneration and regeneration that follows the pace and pattern of revascularization (Fig. 6.5). The first ingrowing capillaries bring with them inflammatory cells (neutrophils and macrophages) that both begin to remove necrotic muscle fiber debris and stimulate the activation of satellite cells (for details see p. 127). Next, the satellite cells proliferate and fuse to form regenerating myotubes, which then go on to differentiate into cross-striated regenerated muscle fibers. Because of the gradient set up by revascularization, muscle fibers regenerating at the edge of the lesion may have reached the stage of cross-striations while at the same time the center of the lesion may still consist of non-vascularized necrotic muscle fibers.

The scale of ischemia can range from areas of microdamage to a condition called **Volkmann's contracture.** In this condition, pressure on the major arteries of the upper arm can cause necrosis of the entire group of flexor muscles of the forearm. Another cause of ischemic necrosis is free muscle transplantation, which involves transplanting a small muscle to another location without an accompanying blood supply. In this case, most of the central muscle fibers within the graft die from ischemic necrosis, but then regenerate when blood vessels grow back into the graft. If the graft is too large, **fibrosis** develops in the center of the graft before ingrowing blood vessels can restore muscle fiber regeneration in that area (Fig. 6.6).

Chemical damage. A number of chemical compounds are known to cause muscle fiber damage. Some, such as barium chloride, are used as experimental tools, but others are used clinically. Among the latter, several local anesthetics are known to be myotoxic, but their effects in laboratory rodents are considerably more severe than they are in humans. Nevertheless, some of these anesthetics, for example, bupivacaine (Marcaine), can damage muscle fibers in humans if they are injected directly into the muscles, as is sometimes done in sports medicine.

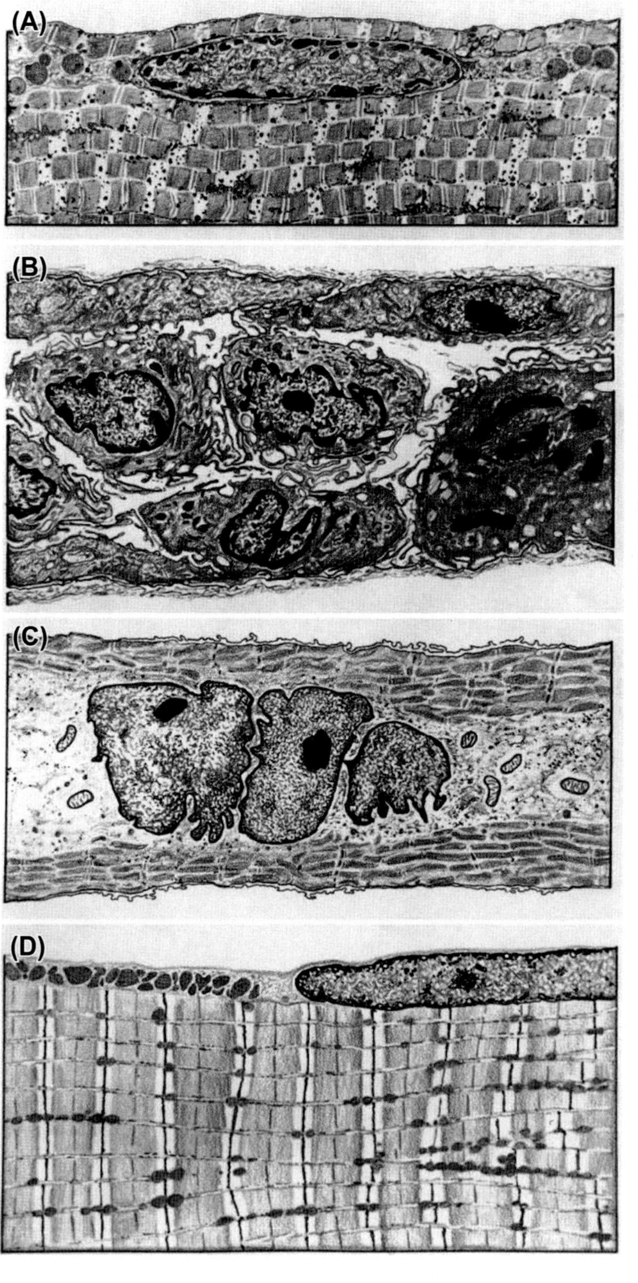

Figure 6.4 Series of drawings showing stages in the degeneration and regeneration of a mammalian skeletal muscle fiber. (A) Initial intrinsic degeneration of an ischemic muscle fiber. The nucleus is becoming pyknotic, and individual sarcomeres are separating because of the action of proteases. (B) Phagocytosis of the degenerating muscle fiber by numerous invading macrophages. Beneath the persisting basal lamina (top) can be seen two spindle-shaped activated satellite cells. (C) Regenerating myotube with central nuclei and newly synthesized bundles of contractile filaments, all beneath the remains of the original basal lamina. (D) Regenerated muscle fiber.

Chemical injections are commonly used to cause muscle fiber destruction in the laboratory because as a rule their toxicity affects muscle fibers directly, but does not harm the microvasculature. As a result, muscle fiber regeneration occurs simultaneously throughout the lesion, rather than along a gradient as is the case in lesions characterized by ischemia.

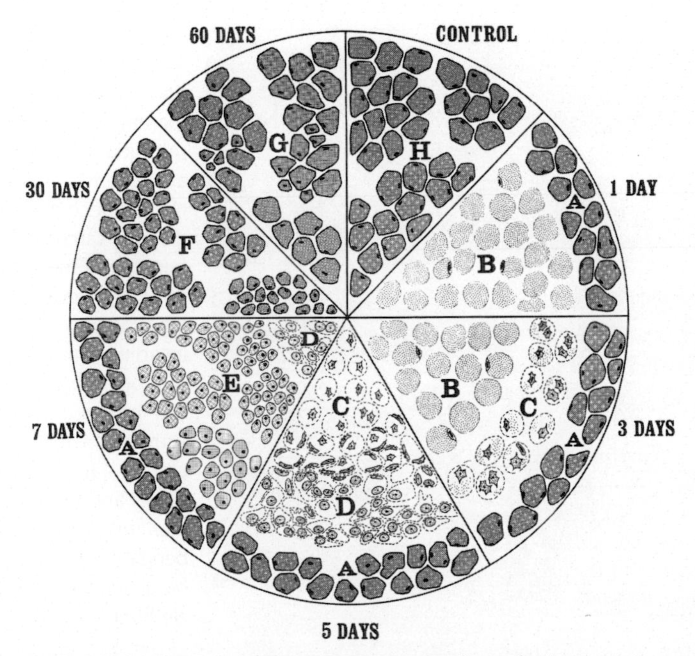

Figure 6.5 Stages of degeneration and regeneration in a freely grafted muscle in a rat. (A) Surviving peripheral muscle fibers that are able to obtain oxygen and nutrients through diffusion; (B) muscle fibers in a state of ischemic necrosis; (C) Degenerating muscle fibers filled with macrophages and activated satellite cells. (D) Early stages of muscle fiber regeneration; (E) early myotubes; (F—H) later stages of muscle fiber regeneration.

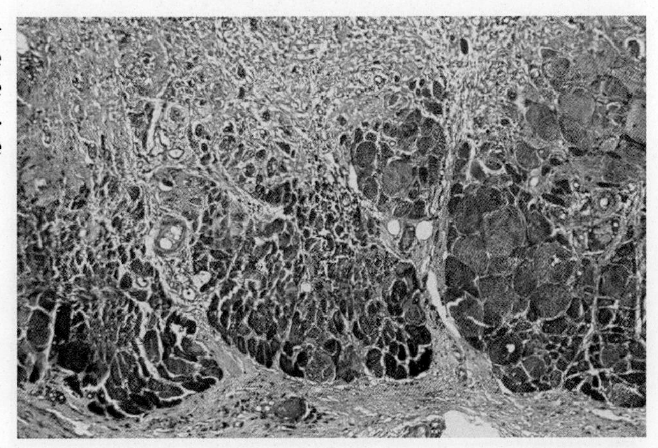

Figure 6.6 Mature graft of a monkey muscle, showing extensive fibrosis (*light blue*) in an area where revascularization was incomplete. Regenerated muscle fibers are stained red.

Muscle fibers can also be damaged by a variety of natural venom myotoxins, especially those found in snake venoms. Of these, notexin (derived from the Australian tiger snake) and cardiotoxin (from a cobra) produce major muscle fiber damage and are commonly used as myotoxic agents in laboratory studies. These venoms contain a phospholipase that lyses the plasma membrane of the muscle fiber. Not only snake venoms, but the venom from certain bees and tarantulas also causes the local breakdown of muscle fibers.

Thermal trauma. Muscle fibers are damaged by extremes of either heat or cold. These conditions, however, also damage other tissue components of muscle, especially the microcirculation. Therefore, the repair process follows a course similar to that seen after ischemia.

Disease-induced muscle fiber degeneration. A number of muscle diseases are characterized by extensive muscle fiber degeneration and subsequent regeneration. These fall into several categories: (1) muscular dystrophies, (2) autoimmune-based inflammatory muscle diseases, (3) metabolic muscle diseases, and (4) congenital myopathies. These are discussed in greater detail in Chapter 7.

The role of muscle fiber damage and inflammation in muscle regeneration

Almost immediately after muscle fiber damage, whether by ischemia or mechanical damage, external Ca^{++} rushes into the damaged muscle fiber and activates intrinsic Ca^{++}-activated proteases (see p. 112), which begin to chemically break down sarcomeric proteins. Products of tissue damage activate mast cells and tissue macrophages, which are present in large numbers in the perimysium and epimysium of normal muscle. Mast cells are primed to respond almost immediately to external stimuli (as they do in acute asthma attacks), and within minutes they release cytokines that both stimulate satellite cells and also call in white blood cells from the circulation. The presence of muscle cell breakdown products, as well as cytokines (e.g., **CC-chemokine ligand 2 [CCL2]**) released by resident tissue macrophages, also stimulates the invasion of blood-borne white blood cells, especially neutrophils, into the area of damage. Muscle damage also activates the **complement system** within the blood, which is yet another stimulus for the local recruitment of white blood cells.

Neutrophils infiltrate the site of damage within hours and reach peak numbers within 12–24 h (Fig. 6.7). These acute inflammatory cells are mildly phagocytic, but they also secrete a large variety of enzymes, which begin to break down products of tissue damage. Cytokines released from acute inflammatory cells attract secondary inflammatory cells from the blood to the site of damage. Neutrophils also produce reactive oxygen species, of importance in promoting degenerative processes. If excessive numbers of neutrophils enter an area of muscle damage, they strongly inhibit muscle regeneration through the production of reactive oxygen species.

During the first day after damage, secretions from neutrophils, as well as resident T-lymphocytes and macrophages, stimulate the immigration of secondary inflammatory cells, namely, monocytes and macrophages from the blood (Fig. 6.8). At least two of these secretions, **interleukin-1** and **tumor necrosis factor-α (TNF-α)**, activate the

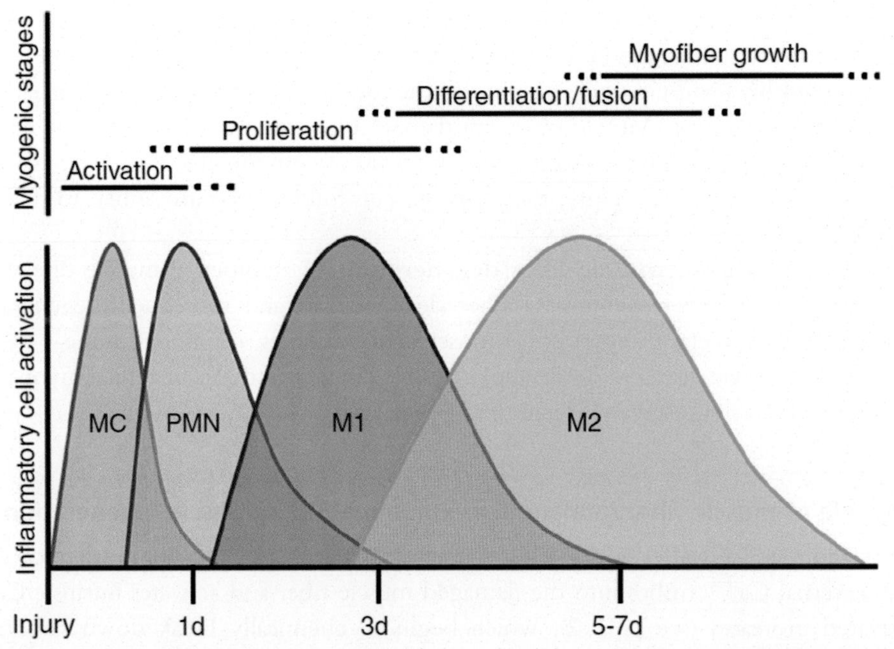

Figure 6.7 Stages of inflammation and myogenesis in a damaged muscle. Mast cells (MC) are among the first cells to become activated. This is followed by the infiltration of neutrophils (polymorphonuclear leukocytes [PMN]), and then proinflammatory (M1) and antiinflammatory (M2) macrophages. *(From Dumont et al. (2015), with permission.)*

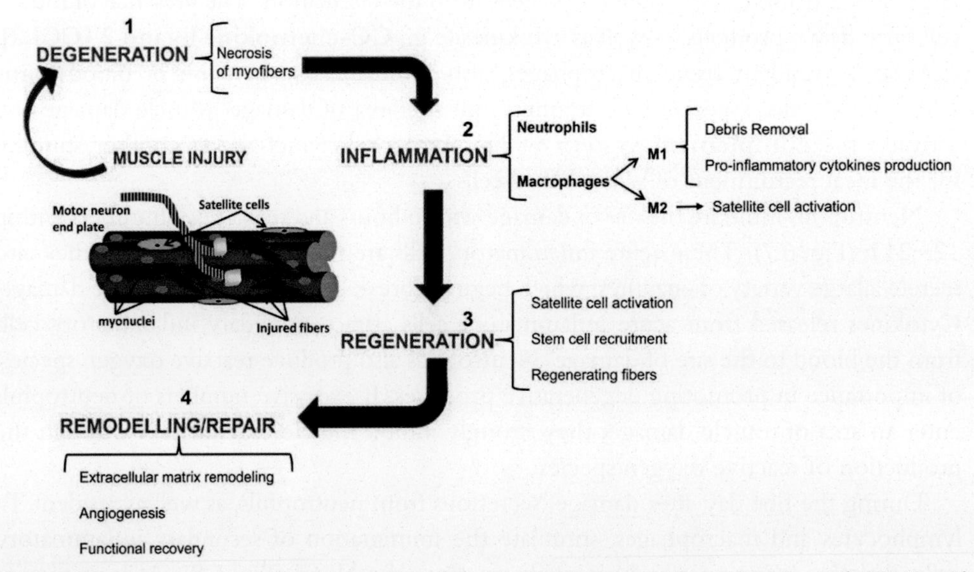

Figure 6.8 Important stages in the process of muscle regeneration. *(From Carosio et al. (2011), with permission.)*

macrophages to a specific pro-inflammatory type (**M1 macrophages**). These macrophages voraciously phagocytize the bulk of the debris resulting from the breakdown of the muscle fibers, as well as dying neutrophils. In addition, the secretions of M1 pro-inflammatory macrophages (interferon-γ, IGF-1, interleukins [IL-6], and TNF-α) stimulate the early activation of surviving satellite cells by stimulating mitosis and increasing their mobility. By 2—3 days in rodents, the persisting basal laminae of the original muscle fibers are filled with phagocytic macrophages. Immediately beneath the basal lamina are spindle-shaped activated satellite cells (see Fig. 6.4B).

Upon phagocytosis of cellular debris, many of the M1 macrophages undergo a transition to M2 (antiinflammatory) macrophages. These cells secrete growth factors, such as IGF-1, TGF-β, and interleukin-10 (IL-10), which stimulate the resolution of the inflammation and also promote the fusion of activated satellite cells and their differentiation into myotubes.

Activation of satellite cells is well underway as the macrophage-dominated late phase of inflammation begins to recede (see below). This inflammatory phase becomes prominent during the second day following muscle injury and is largely resolved by the end of the fourth day, when myotube formation is well underway.

Activation of satellite cells

In normal healthy muscle, satellite cells are kept in a state of quiescence through a variety of molecular mechanisms (see p. 61). After muscle damage, the balance is tipped from promoting quiescence to one promoting activation. Responding to signals from damaged muscle fibers and the inflammatory cells that invade the area of damage, the formerly quiescent satellite cells quickly gear up to enter the mitotic cycle and also develop the capacity to become migratory. Activation of satellite cells occurs not only at the immediate site of injury, but anywhere along a muscle fiber that has become injured. The activated satellite cells then migrate toward the injured area.

Signals from a variety of sources converge upon satellite cells to activate them. One important source is the basal laminae that surround the damaged muscle fibers. Basal laminae are known to sequester a variety of growth factors, which can be released to provide a rapid response to environmental changes. Almost immediately after damage, muscles release **nitric oxide** (NO). NO stimulates the activity of matrix metalloproteinases (MMPs), which act on the basal laminae to release both hepatocyte growth factor (HGF) and fibroblast growth factor (FGF). Both of these growth factors, along with products of muscle fiber damage, stimulate activation of satellite cells. Secretions from inflammatory cells, especially M1 macrophages, are also directly involved in activating satellite cells. Among these signals are TNF-α, interleukins and interferon-γ.

Various activation signals shift satellite cells from the quiescent G_0 phase of the cell cycle to the entry G_1 phase, where they remain until the first DNA synthesis (S phase) occurs 14–18 h after injury. One of the earliest markers of activation is the expression of MyoD, which when phosphorylated in this phase of activation, is involved in facilitating DNA replication rather than its more familiar role in promoting muscle differentiation. As these activation changes are taking place, other intracellular changes begin to reduce the expression of Pax-7, the principal marker of quiescent satellite cells. Equally important, **Notch** signaling, which is of great importance in maintaining satellite cell quiescence in stable mature muscle (see p. 61), is reduced as the connections between the Notch receptor on the satellite cells and the Delta ligands on the muscle fibers are disrupted during muscle fiber degeneration.

Activation involves many molecular and metabolic pathways, which are themselves regulated posttranscriptionally by miRNAs. As many as 350 different miRNAs are known to be involved in such regulation. With the activation of cell cycle genes and the initiation of mitosis, activated satellite cells are often called **adult myoblasts** or **myogenic precursor cells**, which still express the Pax7 marker of quiescent satellite cells, but also now express the muscle regulatory factors MyoD and Myf5 (Fig. 6.9). In summary, the transition from the quiescent to the active state involves many levels of molecular regulation that include signaling molecules, transcription factors, epigenetic regulators, RNA-binding proteins, and noncoding miRNAs.

Satellite cells activated by injury soon become migratory. A prerequisite for migratory activity is a release from the connections (e.g., integrins and cadherins) that hold them in place on both the muscle fiber and the overlying basal lamina during quiescence. Once liberated from such constraints, activated satellite cells are not only capable of migrating along the surface of an injured muscle fiber, but they are also able to penetrate basal laminae and even endomysial connective tissue during lateral migrations. Such lateral migrations may extend several mm.

Migrating cells need a substrate, and the major substrate in muscle regeneration is the MMP-altered basal laminae that surround the damaged muscle fibers. The laminin component of the basal laminae is of particular importance in facilitating satellite cell migration.

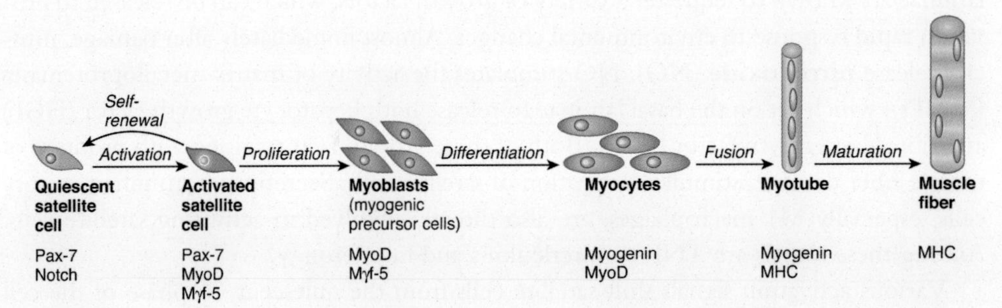

Figure 6.9 Regulatory factors involved in muscle fiber regeneration from satellite cells.

Growth factors and cytokines from several sources act collectively to stimulate the migratory activities of activated satellite cells. One important source is products, for example, the growth factors TGF-β and HGF, coming from degenerating muscle fibers. Another source is M1 macrophages, which release interleukins and TNF-α. A third migratory stimulus is vascular endothelial growth factor (VEGF), which is released from the damaged microvasculature and probably from other sources, as well. All of these attract satellite cells toward areas of muscle damage. In addition, the activation of an **Ephrin/Eph** relationship[1] between activated satellite cells and the healthy portions of muscle fibers provides a form of boundary that pushes the satellite cells away from the healthy portions of the muscle fibers.

This migratory phase brings a population of myogenic cells into the area where regenerative activity is needed. The nature of the migration depends upon the type of initial lesion to the muscle. For small lesions, satellite cell migration is directed from the healthy to the damaged portions of muscle fibers. In ischemic lesions, on the other hand, satellite cell migration follows vascular invasion into the necrotic area. As a rule, migrating satellite cells do not divide, but during their migration the satellite cells are upregulating the cell cycle gene families that will allow them to begin to divide through mitosis.

Regenerative myogenesis

Proliferation of myogenic progenitor cells. Once the satellite cells have been activated and the numerous cell cycle genes have been turned on, these cells undergo a period of intense proliferation in order to produce enough progeny for an effective regenerative response to the injury. Several cycles of proliferation occur before regeneration proceeds to the next stage of fusion of myoblasts and the early differentiation of muscle fibers.

The stimulus for proliferation begins during the early phases of inflammation, and secretions of M1 macrophages, e.g. TNF-α, are mitogenic for satellite cells. Similarly, cytokines, such as the interleukins, that are released from inflammatory cells are mitogenic. Internally, the decline in prominence of the Notch system that takes place during satellite cell activation is accompanied by the upregulating of noncanonical Wnt signaling, which drives proliferation and differentiation in many systems, including regenerating muscle.

The environment associated with dividing myogenic cells is actually inhibitory to the production of collagen by fibroblasts. Under ideal circumstances, the overall balance is tipped toward muscle fiber regeneration; however, under less than ideal conditions, inhibition of fibrosis is less efficient, and scarring becomes dominant.

[1] In many developing systems, closely opposed groups of cells express either Ephrin ligands or Eph receptors on their surfaces. The Ephrin/Eph connection produces a repulsive boundary which keeps the two types of cells from intermingling.

All mitotic divisions of myogenic progenitor cells do not produce the same result. This is important, because during the course of repair, it is necessary to produce not only new muscle fibers, but also sufficient new satellite cells to anticipate needs to repair any future injuries to the muscle. There appear to be several small subpopulations of satellite/myogenic cells that might not follow the same pathways as the main satellite cell population, but at the cellular level the processes outlined below serve as a good example.

Satellite cells can undertake two types of mitotic divisions with respect to the substrate upon which they are located. The difference between these two types of division involves the orientation of the mitotic spindle. In a **symmetrical (planar) division**, the mitotic spindle is oriented parallel to the substrate, and both daughter cells are in contact with the substrate (see Fig. 3.5). After such a division, both daughter cells share the same fate. In an **asymmetric (apicobasal) division**, the mitotic spindle is oriented perpendicularly to the substrate, and typically the two daughter cells follow different pathways of differentiation. If stem cells or precursor cells are involved, the daughter cell abutting the substrate (basal cell) retains it stemness quality, whereas the apical cell goes on to differentiate into a more specialized cell. A key to the different results of such a division is the expression of **Numb**, a Notch inhibitor, in the apical daughter cell. The actions of Numb release the daughter cell from the restraints on differentiation imposed by Notch.

In the case of satellite cells, the basal daughter cell of an asymmetric division retains its stem cell characteristics, including continuing to express Pax-7, and remains in the satellite cell category, whereas the apical daughter cell will become a committed myoblast, expressing MyoD, that will go on to fuse with others to form a regenerating myotube. In a symmetrical division, both daughter cells could either remain as satellite cells or enter the directly myogenic pathway.

Muscle fiber regeneration. At some point, the activated satellite cells withdraw from the mitotic cycle and undertake a program of differentiation to form regenerated multinucleated muscle fibers. This process first involves commitment of individual postmitotic myoblasts to the differentiative process. Following their commitment, the first visible step is the fusion of myoblasts to form multinucleated myotubes.

A major stimulus for the differentiation of proliferating satellite cells is the conversion of the macrophage population from the M1 to the M2 type. M1 macrophages produce proinflammatory cytokines (e.g., TNF-α, IL-6, and interferon-γ), whereas M2 macrophages secrete antiinflammatory cytokines, such as transforming growth factor-β (TGF-β) and IL-10. Importantly, M2 macrophages, which congregate around proliferating satellite cells, also stimulate the production of myogenic regulatory factors in these cells.

Commitment to differentiation first requires activation of the myogenic regulatory factor **MyoD** along with **Myf5** (see Fig. 6.9), which is often expressed even in quiescent satellite cells. Among its many functions, MyoD is heavily involved in suppressing the activities of the cell cycling genes, thus pulling committed myoblasts out of the proliferation pool. The change of state during the myogenic commitment of myoblasts is reflected in the downregulation of the satellite cell markers Pax-7 and later Myf5. Heavily involved in this change of state are a number of miRNAs, which play important roles as modulators of the expression of new genes, for example, MyoD, or the repression of others, for example, Pax-7 and cell cycle genes. Canonical Wnt signaling is a major driver of the differentiative phase of regenerative myogenesis.

The next step in regenerative myogenesis is the fusion of committed myoblasts either to one another to form myotubes or to persisting portions of injured muscle fibers. The basic elements of fusion during normal muscle growth are covered on p. 64 and Fig. 3.6. There is little evidence to suggest that significant differences exist between that and the fusion that occurs during regeneration.

Upon fusion, the satellite cell, now part of a regenerating myotube, turns its attention to differentiation, namely, producing the intercellular proteins and metabolic pathways that will be needed for mature muscle function. This process, however, is a gradual one. Instead of synthesizing and arranging an adult set of contractile proteins immediately, the regenerating muscle fiber closely recapitulates the process of embryonic muscle fiber formation. One of the early steps in this process is the transition from the early myogenic regulatory factors, MyoD and Myf5 to regulatory factors that enhance differentiation, specifically myogenin and MRF4 (see Fig. 6.9).

Muscle fiber regeneration roughly recapitulates the morphology seen in embryonic muscle fiber development, although there are some significant differences (Fig. 6.10). As with most regenerative processes that take place within the adult body, the regeneration of a muscle fiber is connected with the body in ways that are not present in an embryo. The regeneration of most muscle fibers begins within the basal laminae of degenerated preexisting muscle fibers. Only later in regeneration do the regenerating muscle fibers begin to form their own basal laminae. In many circumstances, regenerating muscle fibers develop in connection with intact motor nerve fiber endings or at least the preserved matrices of the original neuromuscular junctions.

Regenerating myotubes are initially relatively simple structures, defined by one or more rows of central myonuclei and the absence of cross-striations. The sarcoplasm is rich in RNA, and at the histological level, early myotubes exhibit cytoplasmic basophilia,[2] which gradually transitions into eosinophilia as the myotubes fill with contractile

[2] In histological preparations, nucleic acids are negatively charged and stain with basic dyes—therefore the term basophilia. In contrast, most proteins and many carbohydrates are positively charged and are stained by acid dyes. With the common histological stain, hematoxylin and eosin, hematoxylin stains the nucleus a deep purple because of its content of DNA. The cytoplasm of most cells stains pink with eosin because of its protein and carbohydrate content. In many developing cells, hematoxylin stains the cytoplasm with a faint purplish hue, which is a reflection of the staining of densities of cytoplasmic RNA, for example, rough endoplasmic reticulum or aggregations of ribosomes.

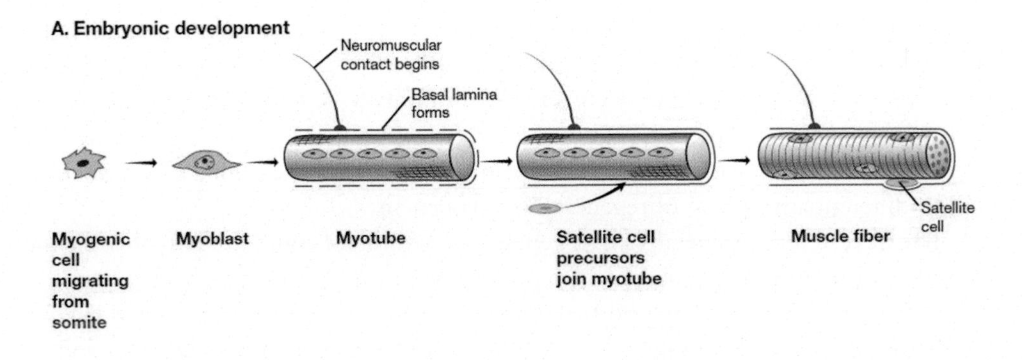

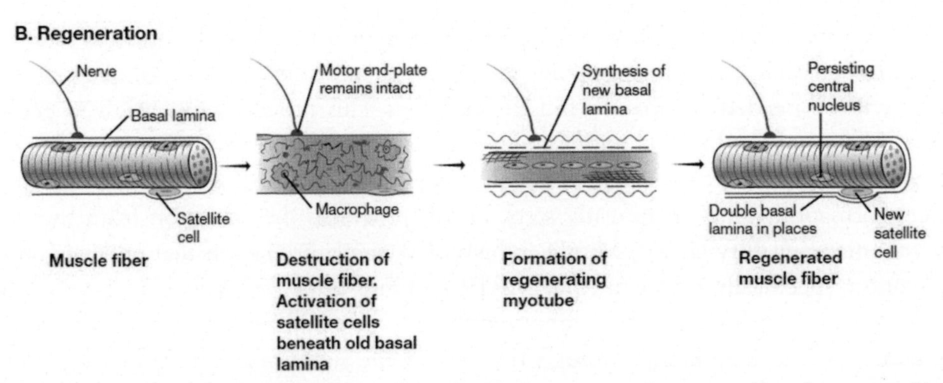

Figure 6.10 Comparison between the embryonic development and regeneration of a muscle fiber. *(From Carlson (2007), with permission.)*

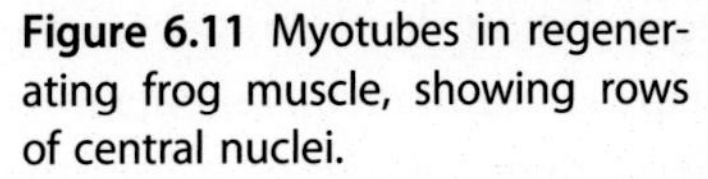
Figure 6.11 Myotubes in regenerating frog muscle, showing rows of central nuclei.

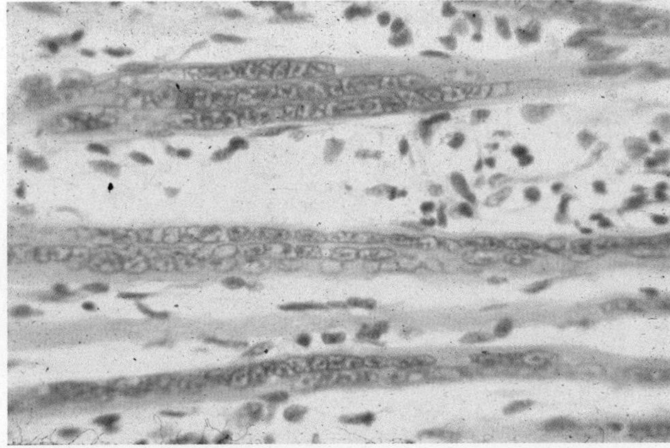

proteins (Fig. 6.11). As contractile proteins are synthesized, bundles of myofibrils begin to accumulate first at the outer edges of the sarcoplasm and then fill in toward the center (see Fig. 6.4C). With the organization of myofibrils, the appearance of cross-striations at the histological level shows that sarcomeres are beginning to form (Fig. 6.12). Over the course of several weeks, the rows of central nuclei begin to break up, and many of the nuclei migrate to the periphery of the regenerating muscle fiber (see Fig. 6.12A). Mature regenerated muscle fibers are filled with myofibrils with well-aligned sarcomeres (see Fig. 6.4D). Even in mature regenerated muscle fibers, many nuclei remain in a central location (Fig. 6.13). The presence of central nuclei is strong evidence in favor of a muscle

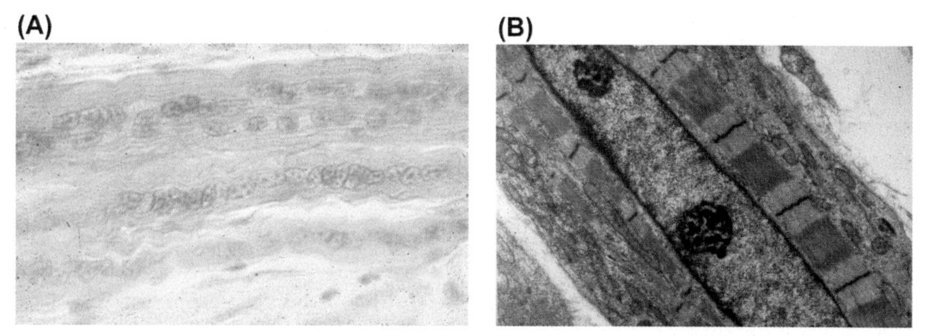

Figure 6.12 (A) Transition between myotubes and early muscle fibers. Central nucleation, seen in the bottom two late myotubes, gives way to migration of the nuclei toward the periphery and the appearance of cross-striations, as seen in the top muscle fiber. (B) Electron micrograph of a rat myotube, showing a central nucleus and newly formed myofibrils with distinct sarcomeric structure.

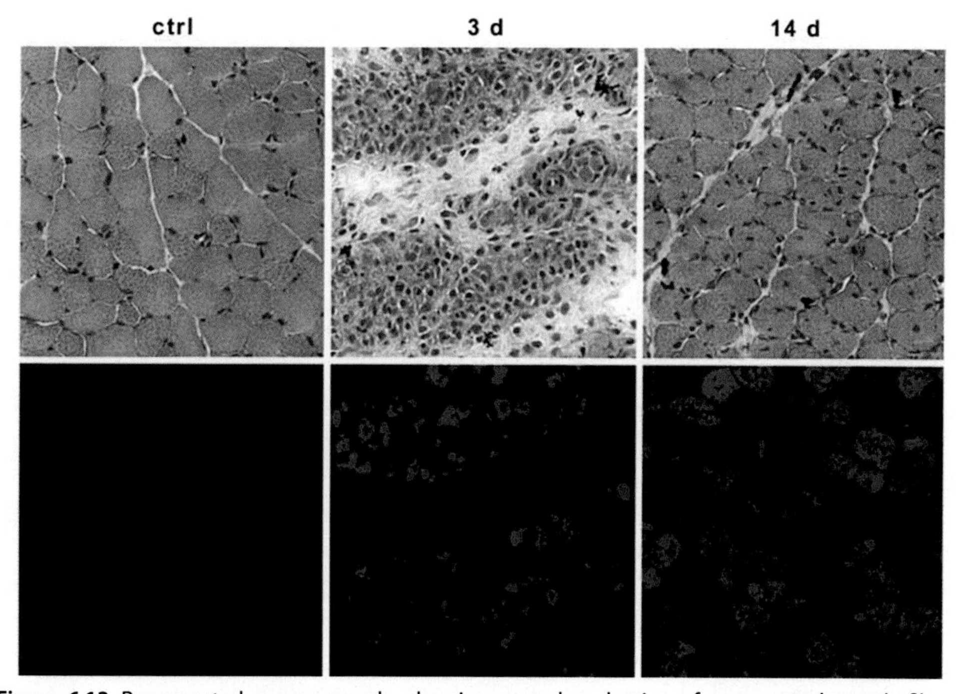

Figure 6.13 Regenerated mouse muscle, showing central nucleation of regenerated muscle fibers. The bottom row shows the expression of embryonic myosin heavy chain in the regenerating muscle fibers. *(From Ciciliot and Schiaffino (2010), with permission.)*

fiber's being a regenerated one. Upon the completion of regeneration, the mature regenerated muscle fibers have accumulated a normal complement of satellite cells, which have pulled out of the mitotic cycle. This ensures the possibility of new regeneration in case of injury to the regenerated muscle. Such cycles of degeneration and repeated regeneration are a feature of many muscle diseases (see Ch. 7).

At the protein level, skeletal muscle fiber regeneration also roughly recapitulates the embryonic pattern of isoform transitions in the expression of contractile proteins. Initially, all regenerating muscle fibers express the sequence of embMHC and then neoMHC. The next transition leads to the expression of fast myosin heavy chains. The switch from embryonic and neonatal heavy chains to fast heavy chains is the expression of a default program that is independent of innervation. On the other hand, the expression of slow myosins requires functional innervation by slow nerve fibers or an electrical stimulation protocol that mimics the normal firing pattern of slow neurons.

Early muscle fiber regeneration does not require the presence of a motor nerve supply, but in the absence of innervation, the muscle fibers undergo slow atrophy and remain functionally fast. If the regenerating muscle fibers become innervated, they rapidly increase in cross-sectional area and differentiate into either fast or slow muscle fibers.

Other interactions during regenerative myogenesis. Muscle fibers do not regenerate in isolation. During the entire process of regeneration, regenerating muscle cells are both influenced by and, in turn, influence neighboring cells and the extracellular matrix. The role of tissue damage and the local inflammatory environment in activating satellite cells to begin the regenerative process has already been described above (p. 125). Also already described has been the role of the basal laminae surrounding the original damaged muscle fibers as both a reservoir for the release of important mitogenic and growth factors and a substrate for the migration of activated satellite cells. Other major players in the regenerative process are elements of the local microcirculation and cells of the local connective tissue (Fig. 6.14).

Especially in ischemic lesions, the sprouting of regenerating capillaries into the ischemic area slightly precedes the regeneration of muscle fibers. The endothelial cells of the ingrowing capillaries secrete several powerful growth factors that stimulate the proliferation of myogenic cells. Among these are HGF, IGF-1, and VEGF. Paracrine interactions between myocytes and endothelial cells provide an environment important for vascular sprouting. In addition to the stimulatory environment provided by the presence of VEGF, interactions between myocytes and endothelial cells sets up the **angiopoioetin-1/Tie-2 receptor** system that is one of the most important stimuli for vascular sprouting. Associated with regenerating capillaries are **pericytes**, which themselves have stem cell-like properties (whether or not these cells have any myogenic potential remains debated). Pericytes influence both early and later stages of regenerative myogenesis through their secretions of IGF-1 and angiopoietin-1.

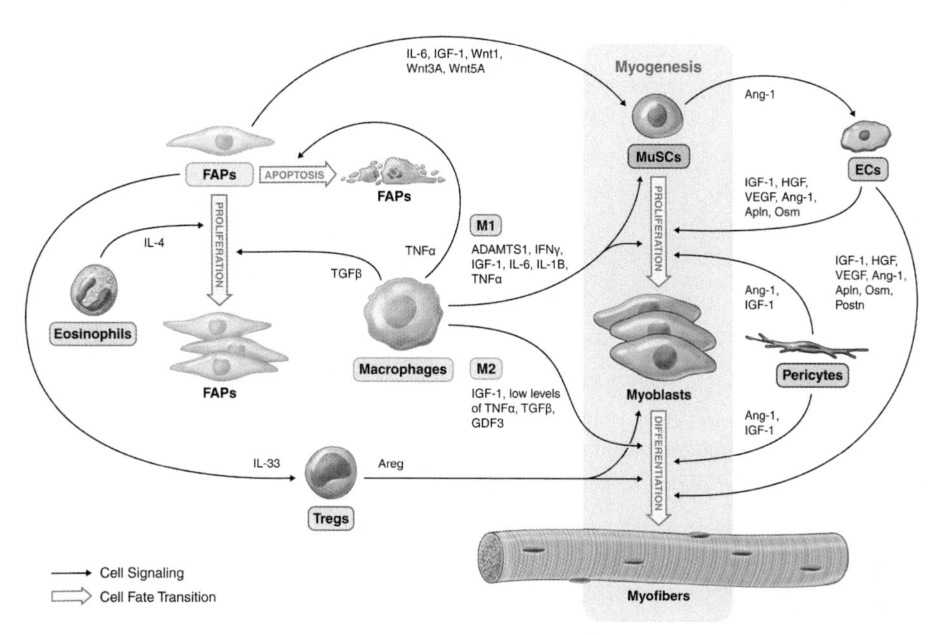

Figure 6.14 The role of other nonmuscle cells and paracrine signaling in muscle regeneration. *Apln*, apelin; *ECs*, endothelial cells; *FAPs*, fibroadipogenic precursors; *GDF*, growth differentiation factor; *IGF*, insulinlike growth factor; *M1* and *M2*, proinflammatory and anti-inflammatory macrophages; *MuSCs*, muscle stem cells; *Osm*, oncostatin N; *Postn*, periostin; *TGF*, transforming growth factor; *TNF*, tumor necrosis factor; *Tregs*, regulatory T cells. *(From Wosczyna and Rando (2018), with permission.)*

One of the prominent types of connective tissue cells in early muscle regeneration is the **fibroadipogenic progenitor cell (FAP).** These cells, whose numbers peak around 3 days postinjury following stimulation by secretions of eosinophils and M1 macrophages, have the capacity to form both fibrous connective tissue or adipose tissue. Stimulated by macrophages and eosinophils, under optimal conditions they secrete a number of factors (e.g., IL-6, IGF-1, Wnts) that support the early stages of muscle fiber regeneration before their numbers recede (see Fig. 6.14). Under less than optimal conditions for regeneration, they can become dominant over muscle fiber regeneration and produce both scar tissue or fat in place of muscle. In healthy regeneration, FAPs are inhibited from differentiating into adipocytes by IL-4 secreted by eosinophils. In normal regeneration, most FAPs undergo apoptosis, induced by TNFα secreted by M1 macrophages. FAP cells can also form cartilage and bone. In regenerating tendinous tissue, they can form well-organized nodules of skeletal tissue (Fig. 6.15).

The influence of innervation upon the character of myosins produced by regenerating myotubes has already been covered. Even in the absence of nerve fibers, the

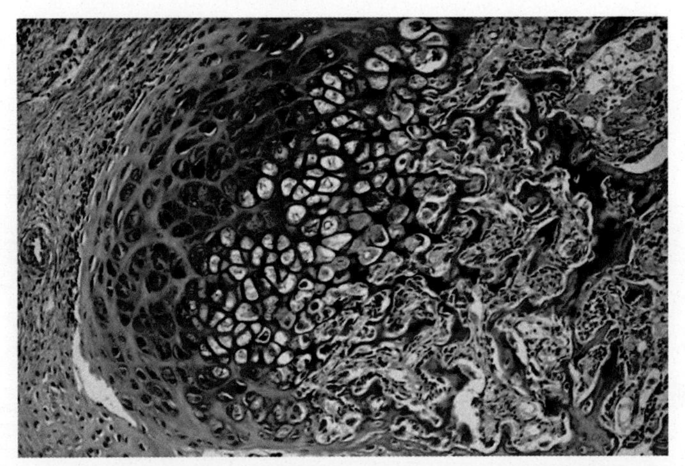

Figure 6.15 Nodule containing bone, cartilage, and bone marrow that formed in a rat minced gastrocnemius muscle regenerate. In addition to these tissue types, the nodule even shows the major structural elements of an epiphyseal plate. *(From Carlson (2007), with permission.)*

specialized basal lamina of the neuromuscular junction stimulates the regenerating muscle fiber to produce postjunctional specializations. Most important, however, the electrical signals provided by innervation stimulate protein synthesis in the regenerating muscle fibers and prevents them from undergoing atrophy.

The differing importance of innervation between the tissue and epimorphic regeneration of skeletal muscle is noteworthy. In tissue regeneration, which can occur in both mammals and salamanders, early stages of muscle regeneration are nerve-independent, and in the absence of innervation muscle fibers can regenerate up to the early cross-striated stage before atrophy sets in. In epimorphic regeneration, however, the absence of nerves does not allow a regeneration blastema to form, and even the earliest stages of muscle regeneration fail to take place.

Gross morphogenesis and internal architecture

Muscle regenerating by the tissue mode does not activate the overall system of morphogenetic control that guides the formation of muscles in epimorphic regenerative processes. Instead, for the most part, both the overall form and internal architecture of a regenerating muscle are strongly influenced by mechanical forces. Under most circumstances and in most experimental models, when muscle fibers are damaged by chemical agents, disease, or ischemia, there is no significant disruption of either the external form or internal architecture of the muscle because the connective tissue components remain largely mechanically undisturbed. In these cases, the fully regenerated muscle is organized very much like its preinjury counterpart.

After trauma, especially when there is loss or significant disruption of muscle tissue, external forces play a significant role in determining the amount and shape of the regenerated muscle. If the distal portion of a rodent muscle is completely removed, regeneration can follow two courses. If the proximal remnant is allowed to remain free, some muscle fiber regeneration will occur at the cut surface but the overall volume of the proximal muscle portion does not change much. On the other hand, if the proximal portion is surgically connected to a functional tendon, the tension applied to the muscle by the tendon causes the regenerating muscle fibers to grow out a significant distance from the cut surface. *In vitro* studies have also clearly shown that muscle fibers differentiate much more fully when they are grown on a substrate upon which tension is applied.

One of the most disruptive experimental models of muscle regeneration is mincing, in which a small muscle is minced into <1 mm^3 fragments. If the fragments are placed into the same muscle bed, they initially become ischemic but then begin to regenerate as new blood vessels grow into the minced mass. In time, a new, but thin, muscle has regenerated. Its form is that of a generic muscle, with tendon connections at either end. Both blood vessels and nerves grow into regenerate. In contrast to epimorphically regenerated muscles, however, minced muscle regenerates are not replicas of the original muscle in either external form or internal architecture (Fig. 6.16A). In fact, a very similar gross model of such a regenerate can be produced by implanting pieces of an inert material in place of a muscle (Fig. 6.16B). Such a "regenerate" is composed mostly of connective tissue, showing the importance of connective tissue as a morphogenetic scaffold for tissue regeneration.

Within days after mincing, fibroblasts growing from the proximal and distal tendon stumps make connections with a regenerating minced muscle. Before tendon connections are made, the orientation of the regenerating myotubes follows that of the basal lamina fragments within the pieces of the mince, and they are oriented in all three planes (Fig. 6.17A). However, after new tendon connections form and tension is applied to the regenerate, most of the regenerating muscle fibers straighten out and become oriented according to the lines of tension within the overall regenerate (Fig. 6.17B). In an extreme example of this phenomenon, if a muscle mince is placed beneath the skin of the abdominal skin, a new button-shaped muscle regenerates (Fig. 6.18). Internally, the regenerated muscle fibers are oriented in all directions in a seemingly random fashion. If directed mechanical tension is applied to such a mince through transplants of tendon pieces, the regenerating muscle becomes pulled out as a tongue-like protrusion (Fig. 6.19), and internally the muscle fibers are oriented parallel to each other. Subsequent experiments conducted in vitro have also demonstrated that developing muscle fibers responding to directed mechanical forces become oriented along lines of applied mechanical tension. In cases of mass trauma to large muscles in humans, the amount of muscle regeneration is often too small for such mechanical influences to be obvious, but internally, the orientation of the regenerating muscle fibers follows the same mechanical rules.

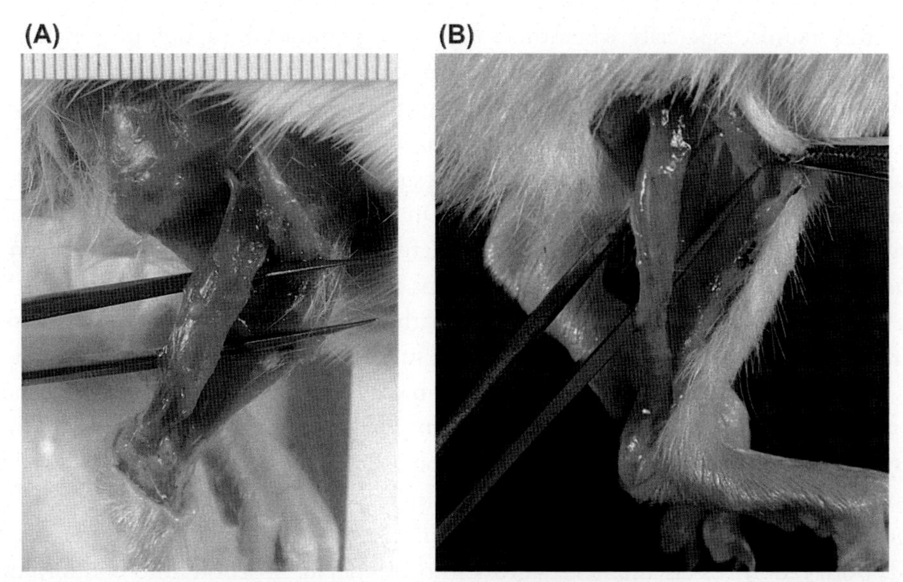

Figure 6.16 (A) A typical regenerate formed from the minced gastrocnemius muscle. Although the regenerated muscle has proximal and distal tendon connections and is vascularized and innervated, its gross shape is that of a generic muscle and does not duplicate that of the normal gastrocnemius. (B) A 13-day "regenerate" formed from implanted bits of Gelfoam (Pharmacia and Upjohn). Here, the gastrocnemius muscle was removed and pieces of the surgical sponge material, Gelfoam, were implanted instead of a muscle mince. The Gelfoam had been molded into the shape of a generic muscle, like that in *A*, with proximal and distal tendon connections. It was vascularized, and the sural nerve (*arrow*) can be seen running over its surface. Despite its similarity to a minced muscle regenerate, this "Gelfoam regenerate" contained no muscle fibers. *(From Carlson (2007), with permission.)*

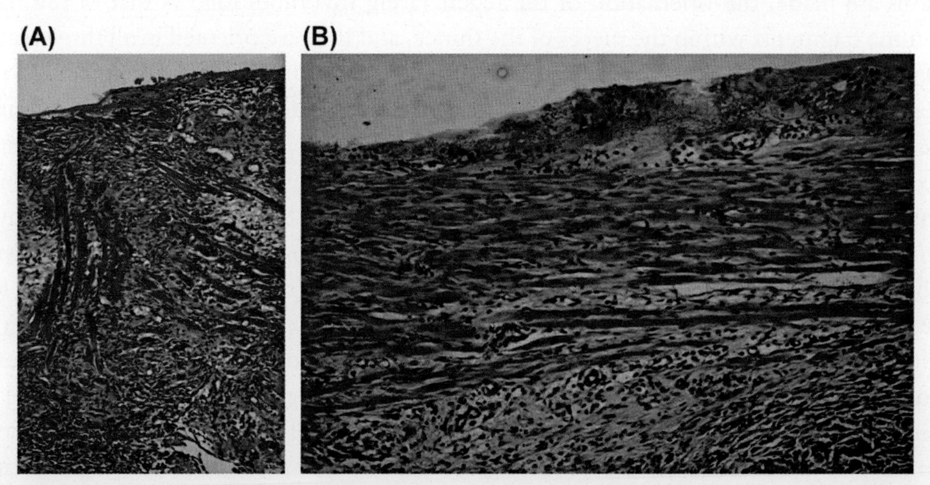

Figure 6.17 The reorientation of regenerating muscle fibers after mincing the gastrocnemius muscle of a rat. In both A and B, the long axis of the muscle extends from left to right. (A) Five days after mincing, the regenerating myotubes are randomly oriented according to the original disposition of the implanted minced pieces of the muscle. (B) Nine days after mincing, the regenerating muscle fibers at the periphery of the mince are becoming oriented parallel to the long axis (and lines of mechanical tension) of the regenerate. *(From Carlson (2007), with permission.)*

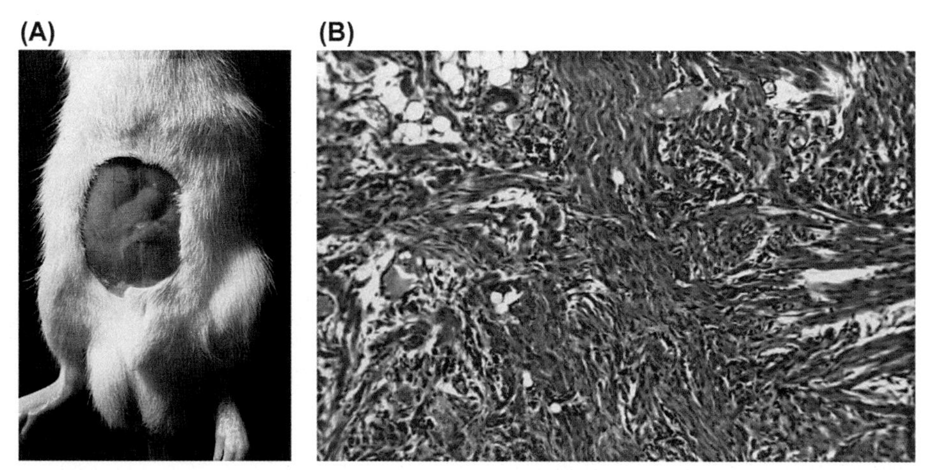

Figure 6.18 Regeneration of minced muscle fragments from the gastrocnemius muscle placed under the abdominal skin of a rat. (A) Fourteen days after mincing, the regenerating muscle (*arrow*) has rounded up into a button-shaped mass. (B) Microscopic section through the same regenerate. The regenerated muscle fibers are randomly oriented in all three dimensions. *(From Carlson (2007), with permission.)*

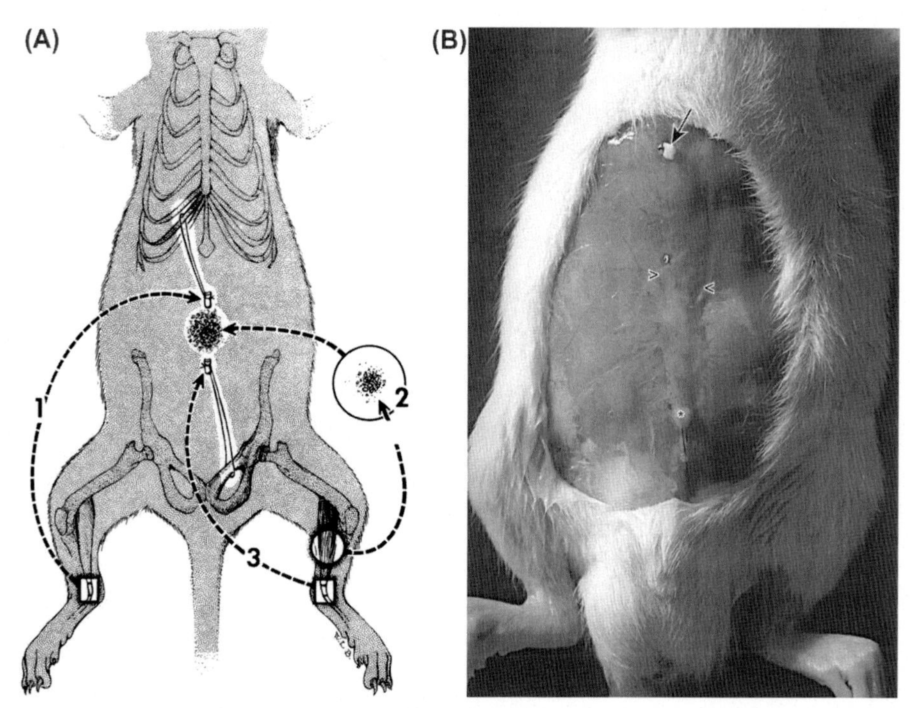

Figure 6.19 Experiment demonstrating the effect of applied tension on the external morphogenesis of regenerating muscle in the rat (Compare with Fig. 6.18, in which tension was not applied.). (A) Half of the gastrocnemius muscle is minced into 1 mm^3 fragments and implanted beneath the abdominal skin. Pieces of Achilles tendon are attached to fixed skeletal points, and the tendons are positioned at the edge of the implanted mince. Within days, the tendon pieces become attached to the implanted muscle fragments through connective tissue. As the animal grows, the tendon stumps are pulled apart, creating continuous tension on the regenerating minced muscle. (B) Gross photograph of a 14-day abdominal mince that regenerated under tension. The cranial tendon piece (*arrow*) broke free, and that end of the regenerate shows radially oriented connective tissue adhesions (*arrowheads*). The caudal tendon piece (*asterisk*) remained connected to the mince and pulled it into a tongue-shaped structure. Within it, the muscle fibers were oriented in a parallel fashion. *(From Carlson (2007), with permission.)*

Functional maturation

Although regenerating muscle fibers become cross-striated within weeks and look mature at the histological level, they are not yet functionally mature. In laboratory rodents, regenerating muscles need at least 60 days to recover normal contractile force, or at least stability. Larger muscles often require up to 6 months to attain functional maturity. Many regenerating muscles never recover their original contractile force. A common reason for this is incomplete reinnervation in the event that there had been some disruption of the motor nerves when the muscle was originally damaged.

Regeneration of muscle spindles

Under conditions of ischemia, mammalian muscle spindles degenerate, but the intrafusal fibers subsequently regenerate. At the cellular level, the sequence of degeneration and regeneration of intrafusal fibers follows the same pattern as that for extrafusal fibers. Although regenerated muscle spindles are not morphologically perfect and are not perfectly innervated, they nevertheless function by discharging impulses in response to stretch.

The relationship between intrafusal fibers and their innervation is complex in both their embryonic development and their regeneration. Whereas sensory innervation is required for the embryonic development of muscle spindles, innervation is not required for the early stages of regeneration in mammals, although the regeneration of muscle spindles in birds appears to be more dependent upon innervation. The persistence of the spindle capsule is important for the support of spindle regeneration in mammals. When the capsule is destroyed, mammalian spindles do not regenerate. In birds, on the other hand, new muscle spindles can regenerate de novo if nonspindle—containing muscles are transplanted into innervated beds of muscles that do normally contain spindles. In the absence of nerves, avian muscle spindles do not regenerate. As is the case with extrafusal fibers, regenerating intrafusal fibers produce fast myosin by default, whereas innervation is required for the expression of slow myosin.

References

Almada AE, Wagers AJ. Molecular circuitry of stem cell fate in skeletal muscle regeneration, ageing and disease. Nat Rev Mol Cell Biol 2016;17:267—79.

Baghdadi MB, Tajbakhsh S. Regulation and phylogeny of skeletal muscle regeneration. Dev Biol 2018;433: 200—9.

Carlson BM. Relationship between the tissue and epimorphic regeneration of muscle. Am Zool 1970;10: 175—86.

Carlson BM. The regeneration of minced muscles. Basel: S. Karger; 1972. p. 128.

Carlson BM. The regeneration of skeletal muscle — a review. Am J Anat 1973;137:119—50.

Chazaud B, Brigitte M, Yacoub-Youssef H, Arnold L, Gherardi R, Sonnet C, Lafuste P, Chretien F. Dual and beneficial roles of macrophages during skeletal muscle regeneration. Exerc Sport Sci Rev 2008;37: 18—22.

Conboy IM, Rando TA. The regulation of Notch signaling controls satellite cell activation and cell fate determination in postnatal myogenesis. Dev Cell 2002;3:397–409.
Dayanidhi S, Lieber RL. Skeletal muscle satellite cells: mediators of muscle growth during development and implications for developmental disorders. Muscle Nerve 2014;50:723–32.
Dumont NA, Bentzinger CF, Sincennes M-C, Rudnicki MA. Satellite cells and skeletal muscle regeneration. Comp Physiol 2015a;5:1027–59.
Dumont NA, Wang YX, Rudnicki MA. Intrinsic and extrinsic mechanisms regulating satellite cell function. Development 2015b;142:1572–81.
Kuang S, Kuroda K, Le Grand F, Rudnicki MA. Asymmetric self-renewal and commitment of satellite stem cells in muscle. Cell 2007;129:999–1010.
Li EW, McKee-Muir OC, Gilbert PM. Cellular biomechanics in skeletal muscle regeneration. Curr Top Dev Biol 2018;1216:125–76.
Maier A. Development and regeneration of muscle spindles in mammals and birds. Int J Dev Biol 1997;41: 1–17.
Mauro A, editor. Muscle regeneration. New York: Raven Press; 1979. p. 560.
Mauro A, Shafiq SA, Milhorat AT. Regeneration of striated muscle, and myogenesis. Amsterdam: Exerpta Medica; 1970. p. 299.
Mourikis P, Tajbakhsh S. Distinct contextual roles for Notch signaling in skeletal muscle stem cells. BMC Dev Biol 2014;14:2.
Qaisar R, Bhaskaran S, Van Remmen H. Muscle fiber diversification during exercise and regeneration. Free Radic Biol Med 2016;98:56–67.
Sampath SC, Sampath SC, Millay DP. Myoblast fusion confusion: the resolution begins. Skeletal Muscle 2018;8:3.
Sass FA, Fuchs M, Pumberger M, Geissler S, Duda GN, Perka C, Schmidt-Bleek K. Immunology guides skeletal muscle regeneration. Int J Mol Sci 2018;19:835–54.
Schmidt M, Schueler SC, Huettner SS, von Eyss B, von Maltzahn. Adult stem cells at work: regenerating skeletal muscle. Cell Mol Life Sci 2019;76:2559–70.
Schiaffino S, Partridge T, editors. Skeletal muscle repair and regeneration. Dordrecht: Springer; 2008. p. 379.
Tidball JG. Regulation of muscle growth and regeneration by the immune system. Nat Rev Immunol 2017; 17:165–78.
Wang H, Simon A. Skeletal muscle dedifferentiation during salamander limb regeneration. Curr Opin Genet Dev 2016;40:108–12.
Wosczyna MN, Rando TA. A muscle stem support group: coordinated cellular responses in muscle regeneration. Dev Cell 2018;46:135–43.
Yin H, Price F, Rudnicki MA. Satellite cells and the muscle stem cell niche. Physiol Rev 2013;93:23–67.
Zammit PS. Function of the myogenic regulatory factors Myf5, MyoD, Myogenin and MRF4 in skeletal muscle, satellite cells and regenerative myogenesis. Semin Cell Dev Biol 2017;72:19–32.

CHAPTER 7

Muscle disorders

Most of us go through life experiencing little more than minor soreness or trauma to our muscles, but a small number of people are greatly debilitated by either inherited or acquired disorders of skeletal muscle. Many of these disorders are genetic, with specific structural or molecular components of muscle fibers targeted by mutations. Others are immunological in nature and are characterized by the infiltration of muscle tissue by immune cells. Still others, often rare, involve the action of certain toxins upon muscle fibers or the neuromuscular junction (NMJ).

This is not a textbook of pathology or medicine, so disorders of muscle will not be presented by specific genetic defects or clinical symptoms. Rather, this chapter will focus first on disorders of specific components of muscle fibers and then how these disorders become manifest as functional impairments.

Disorders of muscle fibers

Disorders of the neuromuscular junction

The normally functioning NMJ is a highly organized synapse that transmits a neural signal to a muscle fiber across a narrow gap (see p. 7). A disturbance at any point along the chain of signaling events can result in a **myasthenic disorder**. The specific location of the disturbance can be in the axon terminal, the synaptic cleft, or the postsynaptic muscle sole plate (Fig. 7.1). A common presenting symptom of many myasthenias is muscle weakness or fatigue, but in some cases, especially involving toxins, the victim is afflicted by paralysis, which could lead to death.

At the axon terminal side of the NMJ, each of the major ion channels is vulnerable to blockage by a variety of natural toxins. One vulnerable site is the voltage-gated Ca^{++} channels. These channels allow the ingress of calcium ions, which are necessary for the fusion of synaptic vesicles to the plasma membrane of the nerve terminal and the subsequent release of acetylcholine. These channels can be blocked by autoantibodies to channel proteins, the basis for **Lambert-Eaton syndrome**, or by **ω-conotoxins**, which are produced by certain species of molluscs. K^+ channels are blocked by **dendrotoxin**, a component of the venom of the mamba snake found in Africa. Na^+ channels can also be blocked by toxins, in this case **tetrodotoxin** and **saxitoxin**. Exposure to both of these toxins can result from eating a variety of marine fish and molluscs, but the actual toxins are produced by algae and are concentrated in their livers or other organs.

Muscle Biology
ISBN 978-0-12-820278-4, https://doi.org/10.1016/B978-0-12-820278-4.00001-3

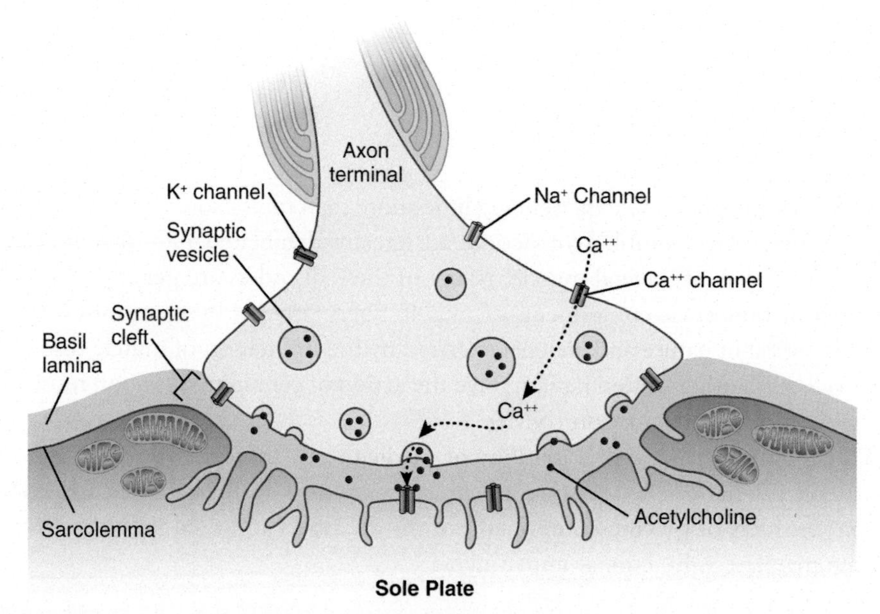

Figure 7.1 Neuromuscular junction, showing important components involved in the transmission of an impulse from nerve terminal to muscle fiber.

A genetic defect in the *CHAT* gene interferes with the function of choline acetyltransferase, resulting in deficient synthesis of acetylcholine. Two devastating toxins can interfere with release of acetylcholine from the synaptic vesicles by cleaving proteins of the membranes surrounding the vesicles. Both are products of bacteria. **Botulinum toxin** is produced by *Clostridium botulinum* and is the basis for the most severe form of food poisoning, usually associated with improper processing of foods. Botulism initially affects the autonomic nervous system, but as it progresses, NMJs are also affected. The roughly 20% mortality of this condition is often due to paralysis of respiratory muscles. **Tetanus toxin**, produced by *Clostridium tetani*, typically enters the body through puncture wounds. In unvaccinated individuals, it attacks peripheral nerves and ultimately causes muscle spasms (e.g., lockjaw) and respiratory paralysis. On the other side of the coin, **α-latrotoxin**, produced by the black widow spider, causes the formation of membrane pores that allow the ingress of Ca^{++} into the nerve terminal. This results in a massive release of acetylcholine from the nerve terminals that can cause muscle fasciculations and pain.

The basal lamina that is interposed between the nerve terminal and the postsynaptic sole plate of the muscle fiber is known to be affected by a few rare mutations that can affect muscle function, mainly presenting clinically as weakness. One such mutation affects the collagen tail of acetylcholinesterase, which is bound to the synaptic basal lamina. Another mutation targets one of the chains of laminin, a major constituent of basal laminae.

At the postsynaptic side of the NMJ, interference with the function of acetylcholine receptors can cause debilitating symptoms and even death. The most common disease is the autoimmune disorder, **myasthenia gravis**, caused by circulating antibodies to acetylcholine receptors. In keeping with the autoimmune basis of this disease, many patients afflicted with myasthenia gravis also have an enlarged thymus or a thymic tumor. The presenting symptoms of myasthenia gravis are muscle weakness, often of the eye muscles or the muscles involved in swallowing. Other patients exhibit overall muscle weakness. These symptoms are due to a reduction in the number of functioning acetylcholine receptors in the muscle fibers and the resulting lessening of muscle fibers' ability to respond to the acetylcholine released by the motor nerve terminals. Specifically, the reduced number of functional acetylcholine receptors in some muscle fibers results in a reduced endplate potential, sometimes to the extent that the firing threshold for stimulating contraction of a muscle fiber is not met. Why some variants of myasthenia gravis affect principally eye muscles or muscles associated with swallowing is not known, but given the different embryonic controls of several groups of craniofacial muscles from muscles of the trunk, it is reasonable to suspect that these muscles may have slight antigenic differences from trunk muscle fibers. Lambert-Eaton syndrome (see above), produces clinical symptoms similar to those of myasthenia gravis, but in this case the cause is a reduction in the quanta of acetylcholine released by the nerve terminals. In effect, the result is the same, because in this syndrome endplate potentials also do not rise to the level of stimulating contraction of the muscle fiber.

Two powerful toxins can cause devastating effects by binding to acetylcholine receptors and interfering with the binding of acetylcholine to these same receptors. One is **D-tubocurarine**, the active ingredient of curare—a plant poison used to tip arrows by natives of the Amazon region. D-Tubocurarine is a competitive inhibitor of acetylcholine, and its effects can be counteracted by increasing the supply of acetylcholine at the NMJ. Another toxin is **α-bungarotoxin**, a neurotoxin found in the venom of certain cobras. This toxin irreversibly binds to acetylcholine receptors and is invariably fatal. Although most of the toxins mentioned above are rarely encountered in real life, they have proven to be invaluable aids in our understanding of the details of neuromuscular transmission. When labeled, their binding to subcellular components can be precisely localized. This, correlated with the functional deficiencies that they produce, is a major reason for our understanding of the precise details of neuromuscular transmission.

Disorders of the sarcolemma and endomysium

For many years, **muscular dystrophy** was a catchall term for a group of rare conditions characterized mainly by progressive muscular weakness with various degrees of severity. Some affected the entire body and ultimately led to an early death; others affected only certain parts of the body with relatively mild symptoms. Some forms of muscular dystrophy (e.g., **Duchenne's muscular dystrophy**) affected only one sex (males) and were determined to have a genetic basis; for others a genetic basis was less apparent. Even the fundamental pathology was uncertain. Many decades saw a raging debate about whether Duchenne's muscular dystrophy was fundamentally a disease of muscle or nerve. This debate ended with the identification of the **dystrophin** molecule in muscle fibers and the recognition of abnormalities of this molecule in muscular dystrophy. With the development of greater sophistication in molecular genetic diagnosis and methods, it is now apparent that most forms of muscular dystrophy are due to a large variety of genetic mutations of key structural molecules of the muscle fiber and their connections to the extracellular matrix.

At the heart of most muscular dystrophies is a large complex of molecules that stabilize the sarcolemma and connect internal structural elements of a muscle fiber to the basal lamina and endomysium. Certain molecules of the extracellular matrix itself are targets of mutations that result in some forms of muscular dystrophy. A few other forms of muscular dystrophy involve mutations of nuclear envelope proteins, sarcomeric proteins, or proteins of the sarcoplasmic reticulum. A common denominator in all of these conditions is the production of muscle weakness, despite differences in the underlying pathology.

The network of relevant protein molecules is often called the **dystrophin–glycoprotein complex** (DGC, Fig. 7.2). These molecules function in a highly integrated fashion, and a defect in even one of these proteins often interferes with the function of the entire complex. Thus, many mutants of single components of the DGC often produce similar clinical symptoms.

The DGC encompasses three domains—the intracellular, the sarcolemma, and the extracellular. The dominant component of the intracellular domain is dystrophin, a 426 kd protein. At one end, the dystrophin molecule links to the subsarcolemmal actin network (not the actin of the thin filaments). A middle part of the dystrophin protein links to proteins of the costameres at the Z-line, as well as to the microtubule network within the muscle fiber. At the sarcolemma, dystrophin connects with the transmembrane proteins **dystroglycans** and possibly, the **sarcoglycans**. At its other end, dystrophin connects with molecules that are involved with signal transduction involving mechanoreception and nitric oxide release. These molecules include dystrobrevin, syntrophins, and nNOS (neuronal nitric oxide synthetase). Because of its immense size and many molecular connections, genetic defects of dystrophin often result in some of the most devastating muscular dystrophies.

Dystrophin and the intracellular domain. The classic disease involving dystrophin abnormalities is **Duchenne muscular dystrophy**. This disease, found only in males, is an X-linked recessive condition that becomes manifest when very young boys first exhibit weakness of the proximal limb muscles. As the weakness progresses, affected boys find themselves unable to rise to a vertical position unaided. They begin to assist the straightening-up process by supporting their upper body by placing their

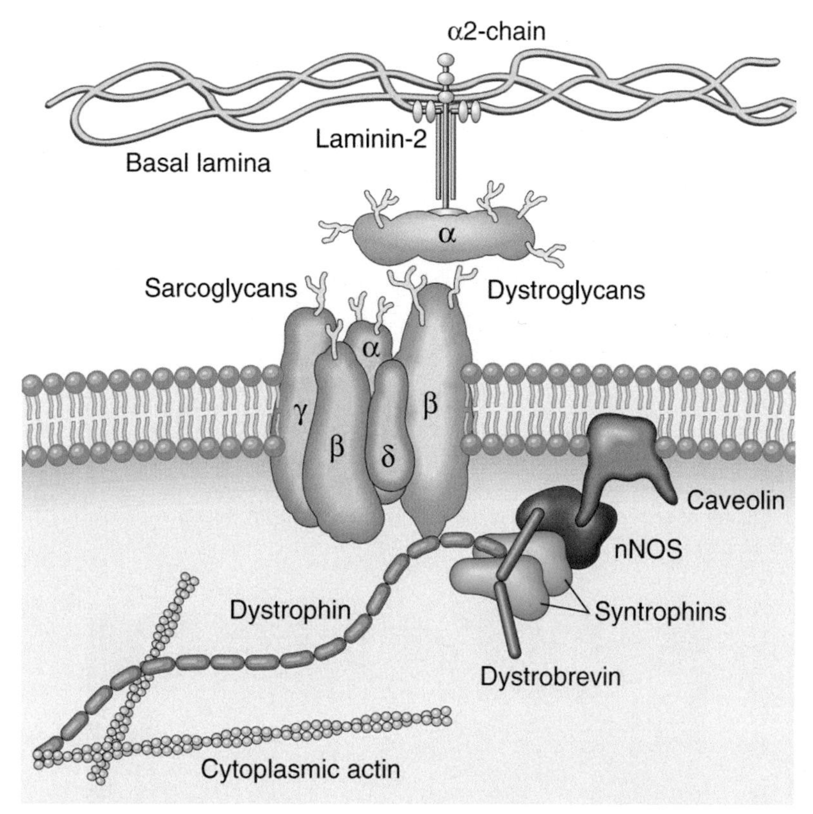

Figure 7.2 The dystrophin–glycoprotein complex. *(From Robbins and Cotran (2015), with permission.)*

hands on the ankles and then progressively walking their hands up their legs until they can finally stand unaided (**Gower's sign**). Early weakness is followed by hypertrophy of certain muscles (e.g., calves and shoulders) while at the same time other muscle groups undergo atrophy. The muscle weakness progresses until most boys are wheelchair-bound by their early teens. Without treatment, patients afflicted by Duchenne's muscular dystrophy often die in their late teens due to respiratory failure or heart problems. A clinically less severe variant is Becker's muscular dystrophy, a condition in which patients may survive until their 30s.

The pathology of Duchenne and Becker muscular dystrophy is characterized by areas of muscle fiber degeneration and regeneration (Fig. 7.3). A characteristic diagnostic feature of Duchenne and most other forms of muscular dystrophy is high blood levels of creatine kinase resulting from the breakdown of muscle fibers. Because of regeneration, many of the muscle fibers are central-nucleated. The cycle of degeneration/regeneration continues for many years until toward the end, muscle fibers become replaced by fibrous connective tissue and/or deposits of fat cells (Fig. 7.4). The underlying basis of the

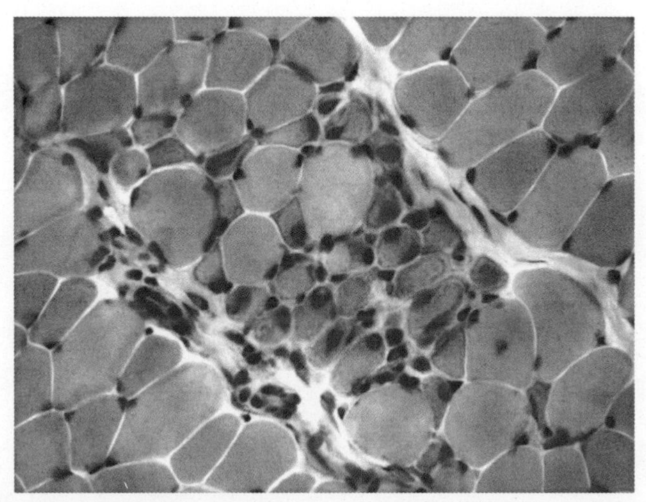

Figure 7.3 Cluster of regenerating muscle fibers in a person with Becker muscular dystrophy. *(From Dubowitz et al. (2013), with permission.)*

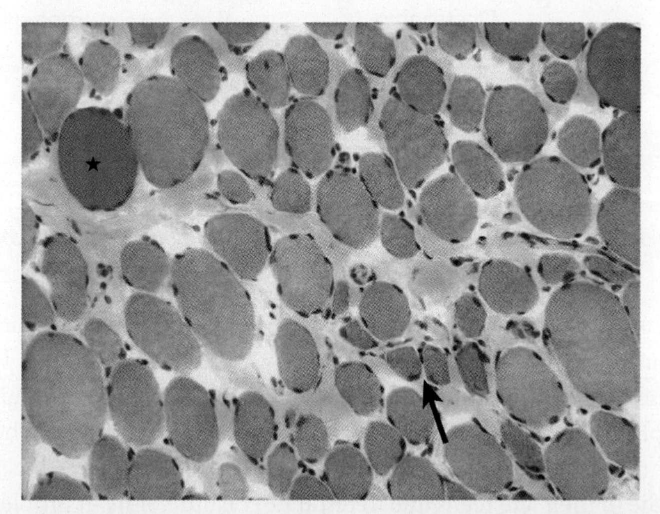

Figure 7.4 A case of Duchenne muscular dystrophy, showing rounded muscle fibers, excess interstitial connective tissue and fat. *Arrow*, basophilic (regenerating?) muscle fibers; *asterisk*, a hypercontracted muscle fiber. *(From Dubowitz et al. (2013), with permission.)*

pathology is a weakening of the sarcolemma due to the absence of functional dystrophin molecules. In the absence of their membrane stabilizing function, muscle fibers cannot withstand the mechanical forces inherent in ordinary muscle contractions (especially eccentric contractions). When mechanically stressed, the dystrophic sarcolemma breaks down, leading to the degeneration of the muscle fiber. After repeated bouts of degeneration/regeneration, the myogenic potential of the muscle stem cells becomes exhausted, and the remaining stem cells begin to produce adipocytes. The basis of endomysial fibrosis in dystrophic muscle remains incompletely understood, but fibrosis is evident even in the early stages of the condition.

The sarcolemmal domain. The sarcolemmal domain of the DGC is dominated by two sets of molecules—the sarcoglycan group and the dystroglycans (see Fig. 7.2). These transmembrane proteins are heavily glycosylated, and mutations involving either the protein core or carbohydrate side chains can be the basis for a number of variants of muscular dystrophy. The **sarcoglycan complex**, which is highly evolutionarily conserved, consists of four tightly associated subunits that are embedded in the sarcolemma. The sarcoglycans are important in maintaining membrane stability through their interactions with both the **dystroglycans** and **dystrobrevin**. A mutation in any of the four subunits is sufficient to weaken the sarcolemma, with the result being forms of muscular dystrophy with a range of severities from mild to severe.

The dystroglycans represent the most important link between the intra- and extracellular domains of the DGC. There are two dystroglycan proteins—α and β. These are encoded by a single gene, but become split posttranslationally. β-Dystroglycan is the transmembrane component of this complex, and as such, it strongly connects with dystrophin. α-Dystroglycan is the extracellular component, and it connects with the basal lamina by providing binding sites for laminin, but at the NMJ, for agrin and perlecan, as well. The associated sugars play an important role in the receptor functions of α-dystroglycan, and mutations affecting either the protein or the carbohydrate components of this molecule can lead to several milder forms of muscular dystrophy by interfering with the connections between the muscle fiber and its surrounding extracellular matrix.

Another component of the sarcolemma is **caveolin-3**, a protein that forms a major part of the caveolae. Caveolae are small vesicles in the sarcolemma that play a prominent role in a variety of signaling events, as well as modulating the release of nitric oxide through its interactions with nNOS. Although not directly connected with the DGC, it plays a minor role in membrane stabilization.

Mutations of caveolin-3, along with the sarcoglycans, can cause mild forms of muscular dystrophy, especially the limb girdle variety. **Limb girdle muscular dystrophy** covers a heterogeneous group of genetically based muscular dystrophies, with severities ranging from mild muscle weakness to the most severe. This group of dystrophies is characterized by weakness and degeneration of muscles of both the pelvic and shoulder girdles, whereas the more distal musculature is relatively spared. In contrast to the dystrophin-related dystrophies, symptoms in limb girdle dystrophies usually do not appear until early adulthood. Another disease associated with mutations in the caveolin-3 gene is **rippling muscle disease**, a hyperexcitability disorder of muscle. Because of the association of caveolae with ion channels within the sarcolemma, the rippling or sometimes bunching response of muscles to percussion or other mechanical stimuli is a logical consequence of this abnormality.

The extracellular matrix. For the most part, the components of the extracellular matrix are not unique to skeletal muscle. Mutations in the genes encoding for two of these components can result in dystrophy-like disease. One is the **laminin α2** subunit of the laminin molecule. Because this molecule is widely distributed beyond muscle, mutations cause symptoms beyond those associated only with muscle. The other ECM component is **collagen-6**, which is strongly represented in skeletal muscle. Collagen-6 is not a fibrillar collagen, but rather it forms micronetworks within connective tissue. Mutations of collagen-6 cause a **collagen-VI myopathy**, sometimes called Ullrich's syndrome. This disease not only affects muscles, with weakness as a major symptom, but also the joints, which can exhibit laxity and later, contractures.

Other disorders of the muscle fiber

Muscle fiber disorders other than the muscular dystrophies are usually based upon inherited defects of organelles other than the dystrophin–glycoprotein complex. Some of these involve structural elements; others are metabolic in nature. As is typical with muscle diseases, most are genetically based.

The nucleus. The nucleus (and the chromosomes it contains) plays an important role in most muscle diseases because they begin with mutations in a variety of genes whose products are critical to normal muscle functioning. A rare form of muscular dystrophy (**Emery-Dreifuss muscular dystrophy**) results from mutations of some of the proteins, such as lamins A and C and emerin, that are located beneath the nuclear envelope. How these abnormalities are translated into a muscular dystrophy phenotype is not understood. This is a common problem with most hereditary muscle diseases. With contemporary molecular genetic technology, the genetic basis of most forms of muscular dystrophy is now known. The clinical symptoms and pathology (the phenotype) are also usually well established. What is missing, however, is an understanding of the developmental pathway leading from the defective gene to the gross phenotype. An excellent example is our inability to understand why some muscular dystrophies affect only certain groups of muscles, for example, those of the limb girdles, proximal versus distal muscles, or certain muscles of the face or neck.

Mitochondria. In addition to their role in normal muscle function, mitochondria are involved in many muscle disorders. This involvement can be at two levels—numbers and morphology of mitochondria or metabolic pathways that are embedded in mitochondria. This section will deal with the former. Metabolic diseases of muscle are covered later in this chapter.

Mitochondria within a skeletal muscle fiber are in a constant state of flux, with new ones created through fission or fusion and old or damaged ones removed by autophagy (mitophagy, see p. 88). Many of the changes at the population level of mitochondria are associated with, rather than causes of muscle diseases. For example, the weakness and reduced endurance seen in Duchenne muscular dystrophy is associated with mitochondrial abnormalities. One minor form of muscular dystrophy is related to a mutation in a gene encoding for one of the mitochondrial membrane proteins.

Mitochondria in skeletal muscle are not homogeneous. They are commonly subdivided into populations of large subsarcolemmal mitochondria and smaller interfibrillar mitochondria, but as many as four morphological types that are all interconnected have been recognized. One prominent condition, **ragged red fibers**, is characterized by the presence of large deposits of subsarcolemmal mitochondria (Fig. 7.5) as a result of proliferation of this subpopulation of mitochondria. This condition, as is the case with some other mitochondrial myopathies, is segmental in nature, meaning that along a single muscle fiber some sections illustrate the pathology, whereas other sections appear to be normal.

Sarcomeric proteins. With rare exceptions (e.g., the gene for titin), mutations of sarcomeric proteins are not primary causes of muscular dystrophies. Nevertheless, a small

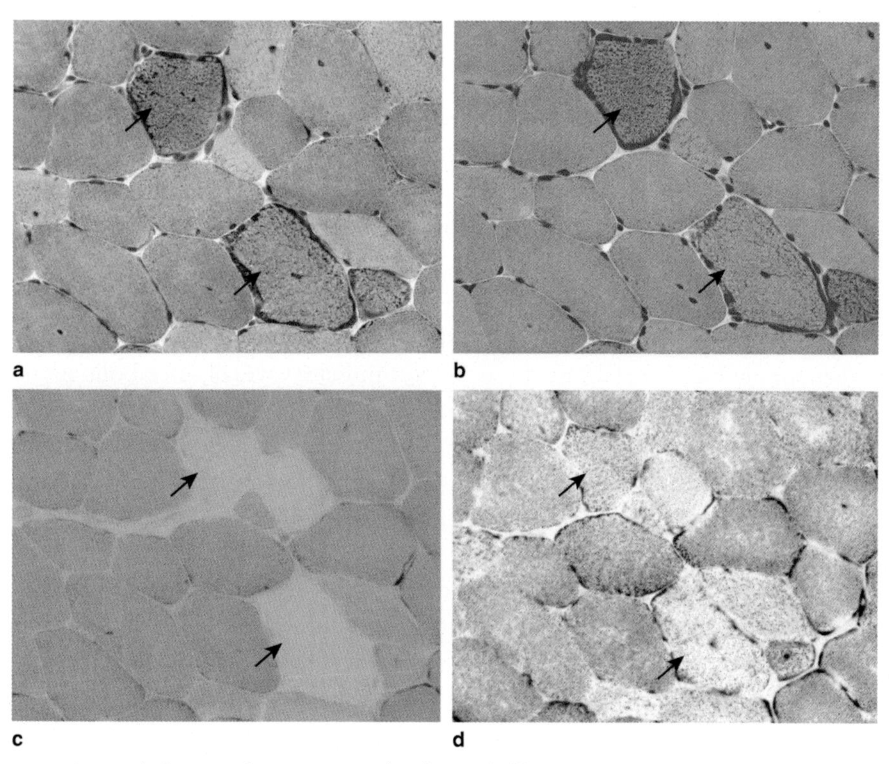

Figure 7.5 A muscle biopsy, showing ragged red muscle fibers (*arrows*), an indicator of mitochondrial dysfunction. *(From Dubowitz et al. (2013), with permission.)*

family of muscle diseases is characterized by the presence of deposits of surplus proteins within the muscle fiber. These deposits are often associated with desmin or thin filaments and are recognizable upon microscopic examination. These myopathies are often mild and nonprogressive, and symptoms, which usually begin in early childhood, consist of weakness or growth disturbances. Even the gene that encodes for α-actin (ACTA1) has been subject to almost 200 described mutations, and mutations of genes for other thin filament proteins (e.g., tropomyosin) have also been described. Genetically based desmin myopathies are progressive and usually begin with weakness of distal muscles, often with associated respiratory insufficiency and cardiomyopathy. Common pathological signs stemming from the over 40 desmin gene mutations are protein accumulations within the muscle fiber.

Ion channel disorders. Mutations affecting the proteins of any of the ion channels (Na^+, K^+, Ca^{++}, or Cl^-) embedded within the sarcolemma can cause a variety of conditions that are characterized by disorders of muscle tone or contractions. Most fall under the general term **myotonia**, which is characterized by a delay in the relaxation of a muscle after voluntary contraction. Specific clinical symptoms depend upon the exact nature of the genetic defect. Dozens of channel protein mutants have been identified.

Other disorders affecting muscles

In addition to disorders, usually genetic or toxin-based, that affect specific components of muscle fibers or their immediate extracellular matrix, skeletal muscle can be severely affected by other conditions. Two of the most prominent types of disorders are metabolic diseases, often genetic and affecting mitochondrial enzymes, and autoimmune disorders leading to inflammatory myopathies.

Metabolic myopathies

Many genetically based metabolic diseases affect muscle. Overall, metabolic myopathies can be divided into three main categories: (1) disorders of glycogen storage and metabolism (**glycogenoses**); (2) mitochondrial myopathies; and (3) disorders of lipid metabolism. Because of their general nature, many of the metabolic myopathies affect a wide variety of other bodily systems. This section will cover only those that affect skeletal muscle in a significant manner.

Glycogenoses. Most glycogen storage diseases are not only confined to skeletal muscle. As might be expected, involvement of the liver and sometimes the heart is often the most prominent pathology. Patients with glycogen storage diseases often present with symptoms of muscle weakness or exercise intolerance. Muscle biopsy is an important diagnostic tool, and at both the light and electron microscopic level, these diseases are characterized by large accumulations of glycogen within the muscle fibers. Two different glycogenoses will serve as examples of this disease category.

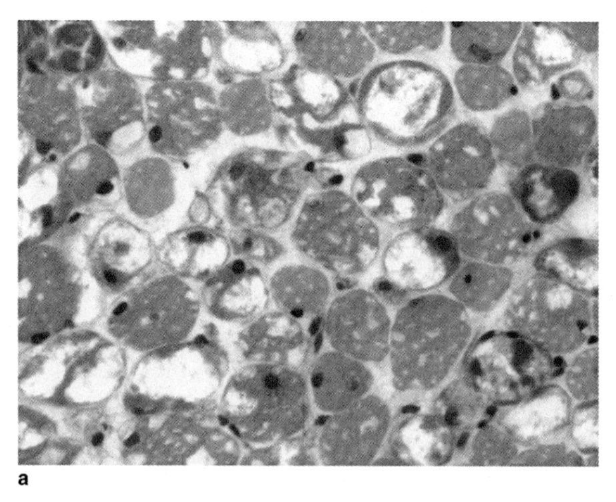

Figure 7.6 Muscle fibers in glycogen storage disease. The empty spaces within the muscle fibers are areas where stored glycogen was dissolved out during preparation of the microscopic preparation. *(From Dubowitz et al. (2013), with permission.)*

The most severe glycogen storage disease affecting skeletal muscle is **Pompe disease**, which is caused by a mutation in the gene for **α-1,4-glucosidase** (lysosomal acid maltase). This disease affects not only skeletal muscle, but also the liver, heart, kidneys, and central nervous system. Affected infants present with severe hypotonia and weakness, frequently associated with heart or respiratory failure and often die within a few years. Skeletal muscle fibers contain massive deposits of glycogen (Fig. 7.6) because the affected enzyme fails to debranch glycogen.

McArdle disease is due to a deficiency in phosphorylase, an enzyme that is critical for the breakdown of glycogen. This disease is unusual in that the pathology is confined to skeletal muscle. The reason for this is that there are three forms of phosphorylase, and the mutation affects only the isozyme (**myophosphorylase**) that is confined to skeletal muscle. Patients with McArdle disease are also afflicted with weakness and especially muscle cramps upon exertion. In general, the clinical onset of McArdle disease occurs later, and the severity is less, than in Pompe disease. A characteristic biochemical feature of McArdle and a number of other glycogen storage diseases is the absence of a rise in blood lactate (a breakdown product of glycogen) after exercise, whereas in a normal subject the rise in blood lactate is three- to fivefold. Although, in general, glycogen deposits are smaller and less pervasive than in Pompe disease, nevertheless they can appear to be prominent upon electron microscopic examination (Fig. 7.7).

A rare glycogen storage disease characterized by glycogen depletion is caused by a deficiency of the enzyme **glycogen synthase**, which is required for the synthesis of glycogen. The absence of glycogen is readily apparent upon histological examination

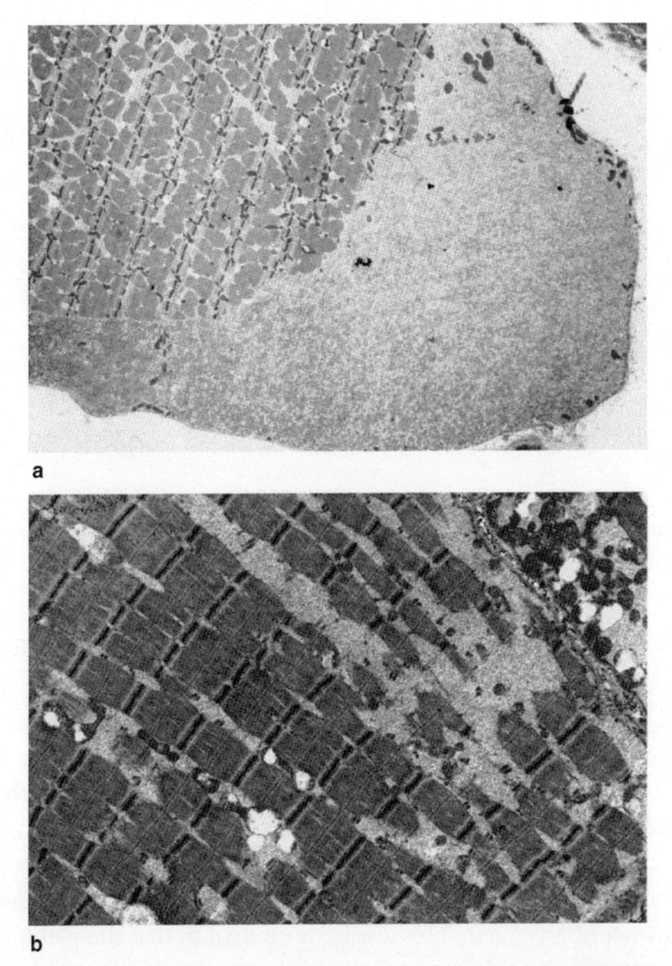

Figure 7.7 Electron micrographs of muscle taken from a biopsy of an individual with McArdle disease. Stores of glycogen (*gray stippling*) are interspersed among the myofibrils. *(From Dubowitz et al. (2013), with permission.)*

(Fig. 7.8). Patients with this condition may have some muscle weakness and shortness of death and are likely to die of cardiac arrest because of the absence of glycogen in cardiac muscle fibers.

Mitochondrial myopathies. Mitochondrial myopathies constitute a large category of diseases whose common element is some sort of disturbance in the mitochondria. Most of these diseases are genetically based and most involve many bodily systems other than muscle. Mitochondrial myopathies include conditions involving substantial changes in numbers of mitochondria, but most are based on abnormalities of individual proteins that are important components of enzymatic pathways.

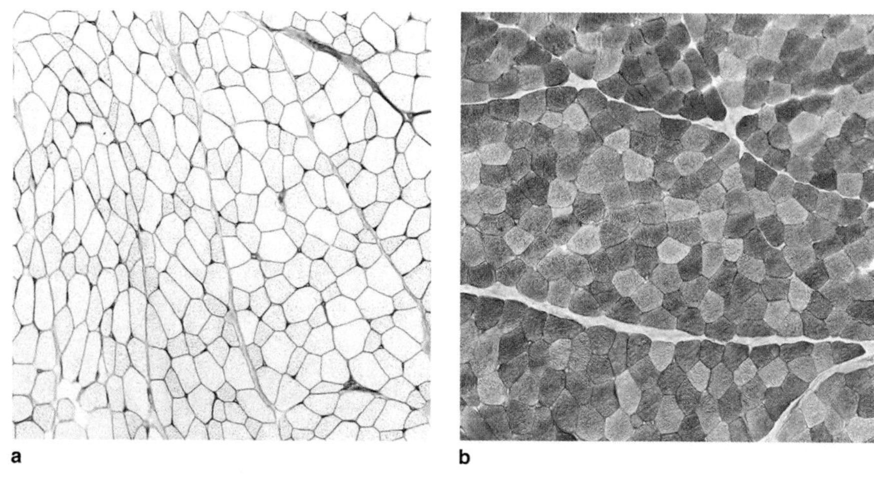

Figure 7.8 (A) Muscle biopsy taken from a 2-year-old girl with glycogen synthase deficiency. (B) Normal control muscle. Both sections were stained with the periodic acid Schiff technique, which stains glycogen deposits violet. Glycogen is virtually absent in the diseased muscle (A). *(From Dubowitz et al. (2013), with permission.)*

Mitochondria are very complex organelles. Presumably derived from the incorporation of symbiotic bacteria into animal cells well over a billion years ago, mitochondria have their own DNA and protein-synthetic apparatus for the production of certain mitochondrial proteins. The mitochondrial genome encodes for 13 of the components of the oxidative phosphorylation system. Most mitochondrial proteins, however, are based upon nuclear DNA and synthesis in the cytoplasm (Fig. 7.9).

A mitochondrion is a double membrane structure with a smooth outer membrane and an inner membrane that is folded into numerous shelf-like cristae (see Fig. 7.10).

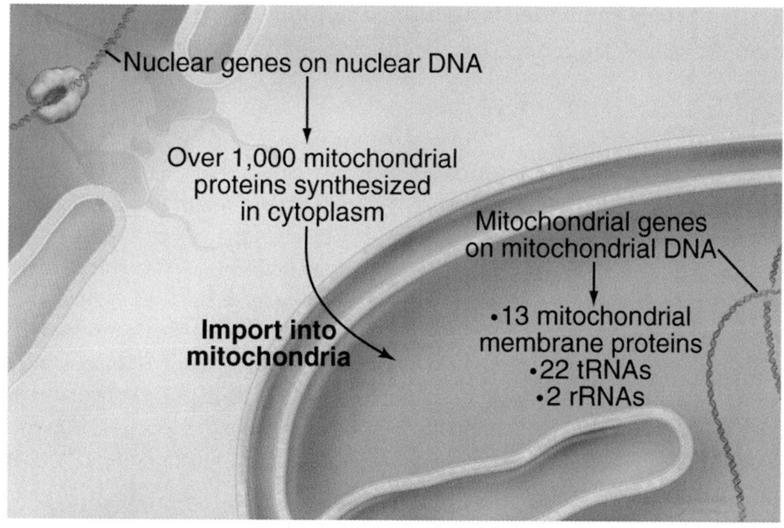

Figure 7.9 Genetic contribution to the synthesis of mitochondrial proteins. *(From Pollard et al. (2017), with permission.)*

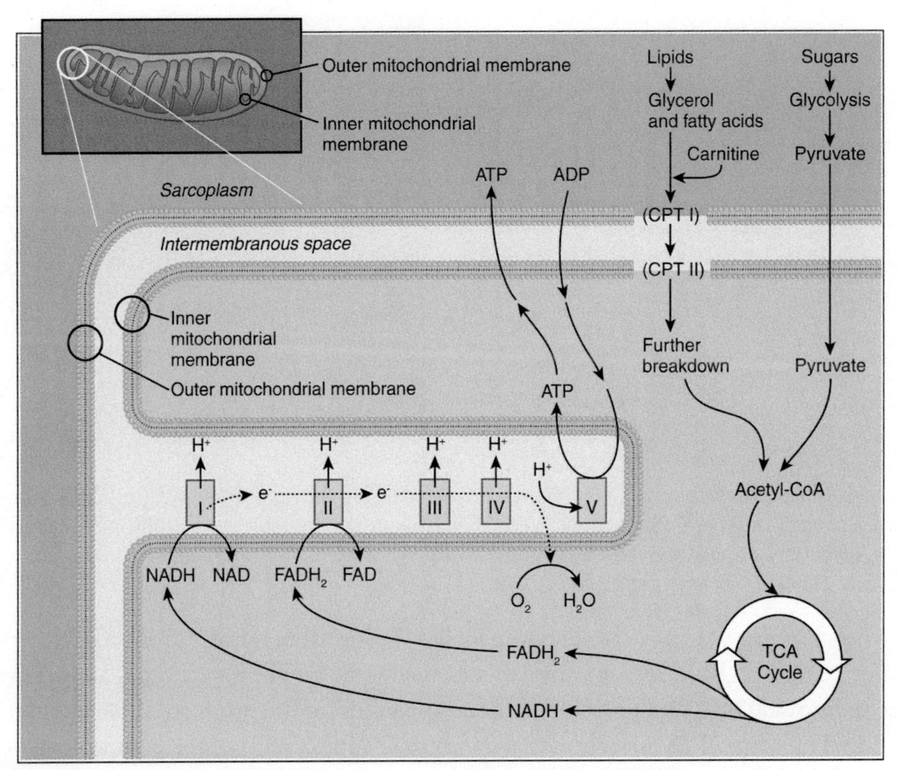

Figure 7.10 Scheme showing major aspects of mitochondrial metabolism in the generation of ATP. *ADP*, adenosine diphosphate; *ATP*, adenosine triphosphate; *CPT*, carnitine palmitoyl transferase; *FAD*, flavin adenine dinucleotide; *NAD*, nicotinamide adenine dinucleotide.

Most of the electron transfer and ATP-generating activities occur on highly registered sequences of enzymes embedded within the inner membrane of the cristae. The overall organization of these sequences has been categorized as five functional complexes (Box 7.1).

BOX 7.1 Mitochondrial enzymes

The oxidative metabolism of carbohydrates and lipids is centered in the mitochondria, which contain important enzyme systems, such as those involved in the TCA cycle and the generation of ATP. Many of these enzymes are physically embedded in the inner mitochondrial membrane, especially on the cristae. In that location, they are highly ordered so that functions like electron transfer can smoothly flow from one enzymatic system to another. This megasystem of enzyme activity has been described as a group of five complexes, each of which is sequentially involved in a specific aspect of the generation of ATP from the breakdown of carbohydrates and lipids (Fig. 7.10). The proteins in these complexes are derived from either nuclear or mitochondrial DNA (see Fig. 7.9). Thirteen of these proteins are encoded in mitochondrial DNA, and over a 1000 are nuclear derived.

BOX 7.1 Mitochondrial enzymes—cont'd

The first four of the complexes in the **electron transport chain** (Table 7.1) accomplish two main functions: (1) moving H^+ obtained from the oxidation of molecules, such as NADH (nicotinamide adenine dinucleotide) to NAD^+ and $FADH_2$ (flavin adenine dinucleotide) to FAD through the inner mitochondrial membrane of the cristae into the space between that and the outer mitochondrial membrane (see Fig. 7.10) and (2) passing the electrons obtained from these reactions down complexes I–IV, where they become involved in the production of water. The hydrogen ions pass through an inner membrane channel in complex V (ATP synthase) and combine with oxygen to become converted to water, whereas electron energy flows into that complex and powers the formation of ATP from ADP.

Table 7.1 Major components of the complexes in the electron transport chain.

Complex	Functional component
I	NADH dehydrogenase
II	Succinate dehydrogenase (part of TCA cycle)
III	Coenzyme Q reductase
IV	Cytochrome c oxidase
V	ATP synthase

Any of the mitochondrial proteins, whether derived from mitochondrial or nuclear DNA, can be abnormal because of mutations, and many of these mutations can result in recognizable disease—of both muscle and other tissues, as well. Mitochondrial-derived mutations are maternally inherited because only maternal mitochondria are passed on from parents to child. On the other hand, mutations of nuclear DNA may show the full spectrum of types of genetic diseases.

Genetically based mitochondrial myopathies can involve specific proteins in any of the five complexes, and a simple gene defect for one protein can inactivate the function of an entire complex. Of the several dozen syndromes that affect oxidative phosphorylation, somewhat more than 50% may demonstrate recognizable muscle pathology. Fewer than that number may exhibit muscle-related symptoms, usually muscle weakness and/or cramps, usually after exercise.

Disorders of lipid metabolism. Another category of metabolic myopathies is disorders of lipid metabolism. Lipid droplets in the form of triglycerides are stored in the sarcoplasm. For full utilization of the energy stored in the sarcoplasmic lipid stores, the neutral fats (triglycerides) must first be broken down into glycerol and fatty acids.

This is accomplished through the action of adipose triglyceride lipase, which catalyzes the first step in this process. If this enzyme is mutated, cytoplasmic muscle lipids do not break down. As a result, the lipid droplets become steadily larger until they occupy a significant proportion of the muscle fiber. With this intracellular change comes progressive muscle weakness. The pathology of this condition is not only confined to muscle fibers because lipid deposits also form in other types of cells, such as those of the liver and central nervous system.

Most steps of lipid metabolism take place in the mitochondria, and several enzymes are involved in transporting lipid metabolites (fatty acids) across both the outer and inner mitochondrial membranes (see Fig. 7.10). In order to pass through these membranes, the free fatty acids must first form complexes with the carrier molecule **carnitine** (β-hydroxy-α-trimethylbutyric acid). Then with the help of two enzymes, the fatty acid—carnitine complex is able to pass through the mitochondrial membranes. Carnitine palmitoyl transferase I (CPT I) is responsible for getting the complex across the outer mitochondrial membrane. It is then passed to CPT II, which takes it through the inner mitochondrial membrane. Once inside the mitochondria, the fatty acids are further broken down in to smaller units (acetyl-CoA), which then enter the citric acid (TCA) cycle for further processing.

Mutations involving any of the enzymes along this pathway can cause muscle pathology, and most affect other organs, as well. Symptoms include weakness as well as pain and tenderness of muscles after exercise, but often patients are minimally symptomatic under normal conditions. Increases in blood CPK levels or myoglobinemia after exercise attest to exercise-induced muscle fiber damage.

Inflammatory myopathies

Not all myopathies are purely genetic. Another category of muscle disease is characterized by the presence of dense infiltrates of inflammatory cells among the muscle fibers. Although a number of inflammatory myopathies have been described, most are very rare. The two most prominent inflammatory myopathies are **polymyositis** and **dermatomyositis**. Muscle involvement in both of these conditions accompanies pathology in other parts of the body, and autoimmunity is a common factor in both although the true cause of neither of these diseases is known.

In both of these conditions, early symptoms consist of weakness and/or aching of axial muscles or proximal muscles of the limbs. Over time, the symptoms may progress to the more distal musculature, as well. By the time symptoms have appeared, the affected muscles show areas of endomysial infiltration by a variety of inflammatory cells (Fig. 7.11). A

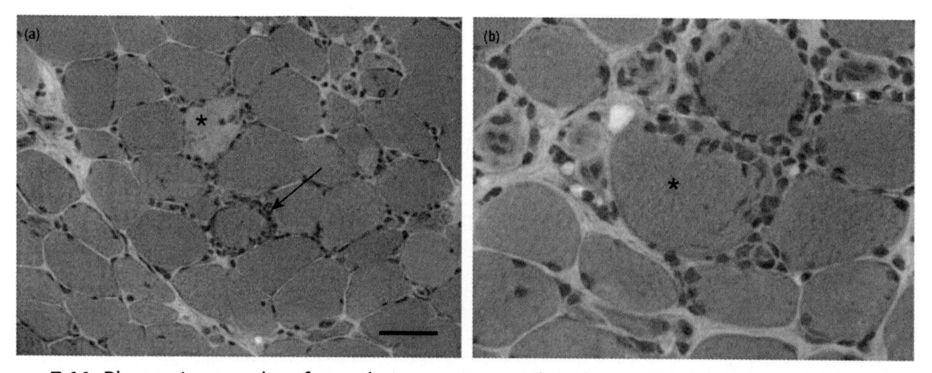

Figure 7.11 Photomicrographs of muscle in a patient with polymyositis. Dense accumulations of inflammatory cells (*arrow*) surround many muscle fibers. *(From Goebel et al. (2013), with permission.)*

common feature of most cases is the presence of MHC-1 (major histocompatibility complex-1) antigen on the sarcolemma of the muscle fibers. Normally, this antigen is expressed only on the blood vessels. Associated with the aggregations of inflammatory cells are muscle fibers in various stages of degeneration and regeneration. An open question is whether the inflammatory cells, especially T-lymphocytes, are a cause or a result of the muscle fiber damage and regeneration. Unfortunately, no pattern of pathology is consistent among all cases, which adds to the frustration of those trying to decipher the underlying causes of these diseases.

Toxic myopathies

Exposure to a large number of agents may lead to toxic myopathies. Many of these agents are drugs; others are toxins. In such myopathies, the dominant pathologies can be either necrosis of muscle fibers or inflammation. A significant contemporary cause of myonecrosis is exposure to **statin** drugs. Taking fibrates in order to lower lipid levels in conjunction with statins is an especially potent combination. Sometimes muscle fiber damage occurs in statin users after intensive bouts of exercise. Common symptoms are pain and/or weakness, which may last for months or years. In extreme cases, products of massive muscle fiber breakdown (**rhabdomyolysis**) may clog the kidneys and cause death. Another important cause of muscle fiber necrosis is heavy drinking, which can cause muscle fiber damage in up to 50% of such drinkers.

An additional response to exposure to toxic agents is inflammatory muscle disease, with symptoms similar to those seen in polymyositis. This can be caused by long-term exposure to interferons for treatment of certain viral conditions, penicillamine treatment for rheumatoid arthritis, exposure to contaminants in dietary supplements containing tryptophan or some animal toxins, such as the fish-related ciguatera poisoning.

Fibrosis

Fibrosis, a large increase in the amount of interstitial connective tissue in regenerating or healing muscle, is a feature of several of the muscular dystrophies, as well as the healing of severely traumatized muscles. A significant question is whether fibrosis is a cause or a consequence of impeded regeneration—or both. Fibrosis can affect many other tissues than muscle, such as the lungs or liver.

Fibrosis is a complex process that centers on the differentiation of a specialized cell type, the **myofibroblast**, from precursor cells, especially **fibroadipocyte precursor cells**. Although many growth factors and cytokines are involved in stimulating the conversion of precursor cells into myofibroblasts, **transforming growth factor-β1** (TGF-β1) plays a central role in this process and serves as a trigger for this conversion. Other important factors are a microRNA (miR-21) and reactive oxygen species, which are products of acute inflammatory reactions. The net result of the activation of myofibroblasts is the accumulation of most components of the extracellular matrix found in the endomysium and perimysium. Among these, the deposition of collagen contributes most to the physical characteristics of fibrotic tissue, which is stiff and limits mobility.

In muscular dystrophies, fibrosis produces thickening of the endomysium and results in the separation of individual muscle fibers from one another (Fig. 7.12). In addition, interstitial fibrosis in muscular dystrophy is associated with the appearance of fat cells among the endomysial connective tissue. After severe mechanical trauma, massive fibrosis typically results in the formation of thick bands of dense collagen which can almost completely limit the function of the injured muscle.

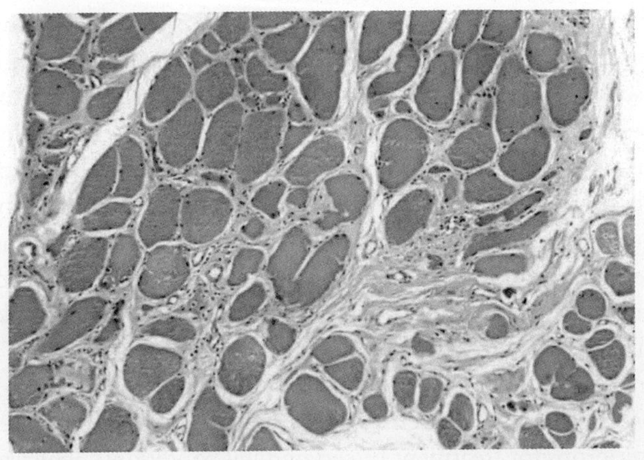

Figure 7.12 Extensive fibrosis in the muscle of a patient with limb-girdle muscular dystrophy. Note the many small atrophic muscle fibers. Inflammatory cells are seen within the interstitial connective tissue. *(From Gea et al. (2006), with permission.)*

References

Bersini S, Giladri M, Mora M, Krol S, Arrigoni C, Candrian C, Zanotti S, Moretti M. Tackling muscle fibrosis: from molecular mechanisms to next generation engineered models to predict drug delivery. Adv Drug Deliv Rev 2018;129:64–77.

Carter JC, Sheehan DW, Prochoroff A, Birnkrant DJ. Muscular dystrophies. Clin Chest Med 2018;39: 377–89.

Dalakas MC. Toxic and drug-induced myopathies. J Neurol Neurosurg Psychiatr 2009;80:832–8.

Dubowitz V, Sewry CA, Oldfors A, editors. Muscle biopsy – a practical approach, 4/e. Amsterdam: Elsevier; 2013. 552 pp.

Gao QQ, McNally EM. The dystrophin complex: structure, function, and implications for therapy. Comprehen Physiol 2015;5:1223–39.

Gawor M, Proszynski TJ. The molecular cross talk of the dystrophin-glycoprotein complex. Ann NY Acad Sci, 1412:62–72.

Goebel HH, Sewry CA, Weller RO. Muscle disease – pathology and genetics. Baltimore: Wiley Blackwell; 2013. 986 pp.

Hilo JA, Olson EN. Muscle – fundamental biology and mechanisms of disease. Amsterdam: Academic Press/Elsevier; 2012. 1408 pp.

Karpati G, Hilton-Jones D, Griggs RC. Disorders of voluntary muscle, vol. 7/e. Cambridge: Cambridge University Press; 2001. 775 pp.

Liewluck T, Milone M. Untangling the complexity of limb-girdle muscular dystrophies. Muscle Nerve 2018;58:167–77.

Mahdy MAA. Skeletal muscle fibrosis: an overview. Cell Tiss Res 2019;375:575–88.

Mercuri E, Boennemann CG, Muntoni F. Muscular dystrophies. Lancet 2019;394:2025–38.

Mohassel P, Foley AR, Boennemann CG. Extracellular matrix-driven congenital muscular dystrophies. Matrix Biol 2018;71-72:188–204.

Sieb JP, Gillisen T. Iatrogenic and toxic myopathies. Muscle Nerve 2003;27:142–56.

CHAPTER 8

The aging of muscle

Many structural and functional properties of skeletal muscle peak in the third and fourth decades of life. Then after a decade or so, during which the muscles can be considered mature and stable, the effects of aging become evident, leading to steady declines in both their mass and contractile power. Aging changes affect not only the muscle fibers, but other components of muscle tissue, as well. Some of the most important changes in muscle fibers are secondary to aging changes in the peripheral nervous system. Both the intramuscular connective tissue and the microcirculation within a muscle also show significant age-related changes. In addition to the muscular system itself, the effects of aging upon the cardiovascular and respiratory systems often directly affect the functional performance of the muscles.

Even during the aging process, muscle retains significant plasticity. When little or not used, muscles in elderly individuals undergo significant atrophy, which can be a reflection of a complex process leading to sarcopenia. On the other hand, aging muscle is also responsive to exercise, even in very old individuals. From a broad functional standpoint, the ability of aging muscle to respond to exercise can mean the difference between an individual's being chair- or bed-bound and being able to participate in what gerontologists call the basic activities of daily living.

Despite its inexorable decline with age, the muscular system has not lost its ability to initiate developmental processes. After injury, aging muscle still regenerates, albeit more slowly and less completely than young muscle. Similarly, aging satellite cells still respond to increased load by participating in a hypertrophic process. At least in rats, occasional new muscle fibers begin to form even in animals approaching death from old age.

As is the case with many other organ systems, it is important to distinguish between changes that are truly age-related and those that are the result of pathology, whether age-related or not. Equally significant are changes due to a sedentary lifestyle, which is common in many older adults.

Broad measures of aging effects on skeletal muscle

In all animals examined to date, from worms to primates, skeletal muscle mass and strength decline with age. By most measures, age-related declines in human muscle structure and function begin to become evident in one's 40s. For the next 10−15 years, the loss of muscle mass is relatively small (0.37%/yr in females and 0.47%/yr in males), but in later life the slope of the decline sharply increases (Fig. 8.1A). The loss in muscle mass with age is not consistent throughout the body. Whereas muscle mass is relatively well

Muscle Biology
ISBN 978-0-12-820278-4, https://doi.org/10.1016/B978-0-12-820278-4.00008-6

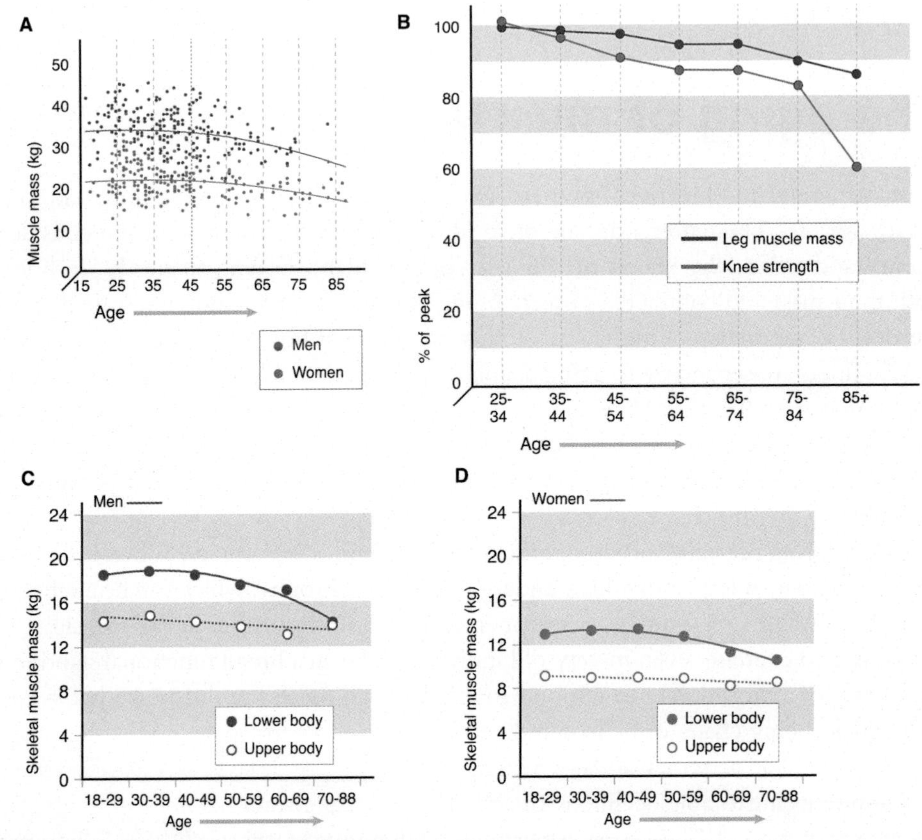

Figure 8.1 (A) Decline in muscle mass with age in men and women. (B) Relative declines in let muscle mass and strength with age. (C) Relative declines in upper and lower body mass with age in men. (D) Relative declines in upper and lower body mass with age in women.

preserved in the upper body, significant declines are seen in lower body muscles especially those of the thigh. At an even more granular level, atrophy of the knee extensors (quadriceps) is greater than that of the knee flexors (hamstrings). Males show a greater rate of decline than females (Fig. 8.1C).

Muscular strength declines at an even faster rate than does mass during aging (Fig. 8.1B). This is in stark contrast to development through maturity, during which the increase in muscle mass and strength is almost perfectly parallel. Age-related functional changes are best reflected in performance measures that involve speed and power. The decline begins at about age 40, and the decline in power is greater than that for endurance performance (Fig. 8.2). As with mass, the decline in power is greater in the lower than in the upper extremities (Fig. 8.3). Measures of gross performance reflect more than simply the condition of the muscles themselves. Changes in cardiovascular and respiratory capabilities must also be factored into the equation.

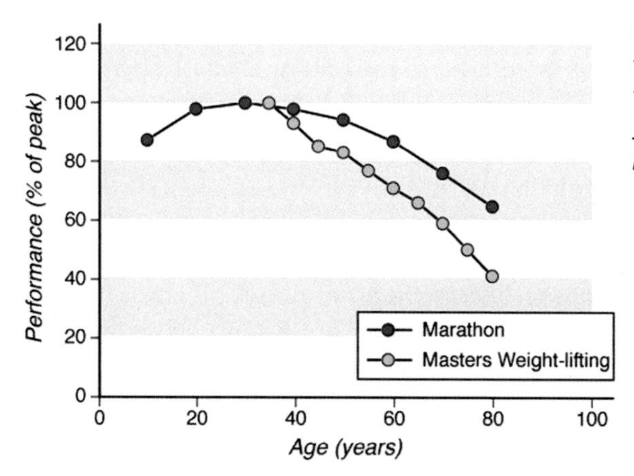

Figure 8.2 Age-related declines in performance of master athletes in the marathon and in weight-lifting (clean and jerk). *(From Carlson (2019), with permission.)*

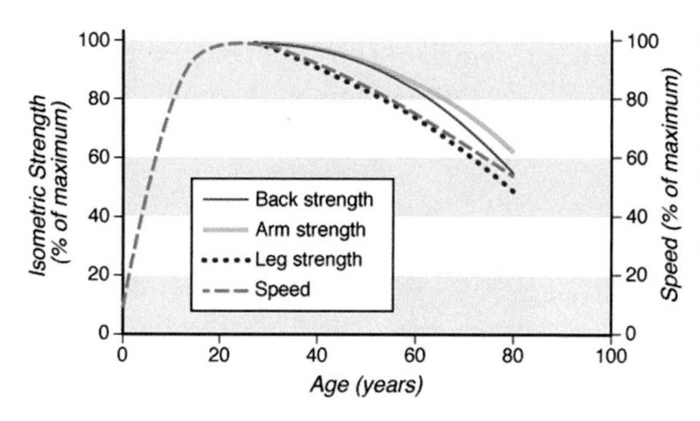

Figure 8.3 Graph illustrating changes in muscle power throughout the human life span. Data on strength are based on a 30-year-old basis of 100%. The speed curve is presented as percentage of maximum speed. *(From Carlson (2019), with permission.)*

The structural basis for muscle loss during aging

The loss of muscle mass during aging is not due to a single factor, but because they make up such a high percentage of any muscle, changes in muscle fibers play a dominant role. Two changes, in particular, are important—the cross-sectional area of individual fibers and the total number of muscle fibers.

The most important contributor to the loss of mass in aging muscle is a reduction in the number of muscle fibers. Because of the size of human muscles, it is not easy to obtain accurate counts of muscle fibers, but several studies have reported decreases in the number of muscle fibers that are proportional to the overall decrease in mass of the muscle. Muscle fiber number decreases only slightly during adulthood, but after age 55—60 significant losses of muscle fibers occur each year. Decreases in muscle fiber number by 25%—40% by age 80 have been reported.

As an individual ages, the average cross-sectional area of muscle fibers has been reported to decrease, but the decrease is not uniform among all types of muscle fibers. Several factors account for this decrease. At the level of individual muscle fibers, the cross-sectional area of Type II fibers decreases from 20% to 50% during the aging process, but the decrease in women is much less than in men. In contrast, Type I fibers change little with age. A second factor leading toward a reduction in average cross-sectional area of all muscle fibers is a change in the proportion of fiber types. With increasing age, the number of Type II (especially Type IIx) in relation to Type I fibers decreases by as much as 40%. Because Type I fibers are thinner than Type II fibers, this tips the scale toward a reduction in mass of an aging muscle. Other characteristics of old muscle are the presence of type grouping of muscle fibers (clusters of muscle fibers of the same histochemical fiber type) and individual muscle fibers containing a mix of myosin types (Fig. 8.4). Despite the above generalization, not all studies would agree with this pattern. Instead they show less change.

According to other human data, average loss of cross-sectional area of muscle fibers is minimal. In fact, according to some studies, the average cross-sectional area of muscle fibers is not only preserved, but may slightly increase. One proposed reason for this is that with the reduction in number of muscle fibers, the remaining fibers undergo some degree of load-induced hypertrophy to compensate for the loss in numbers. Basically, the available data on aging human muscle are not yet sufficient to allow concrete generalizations.

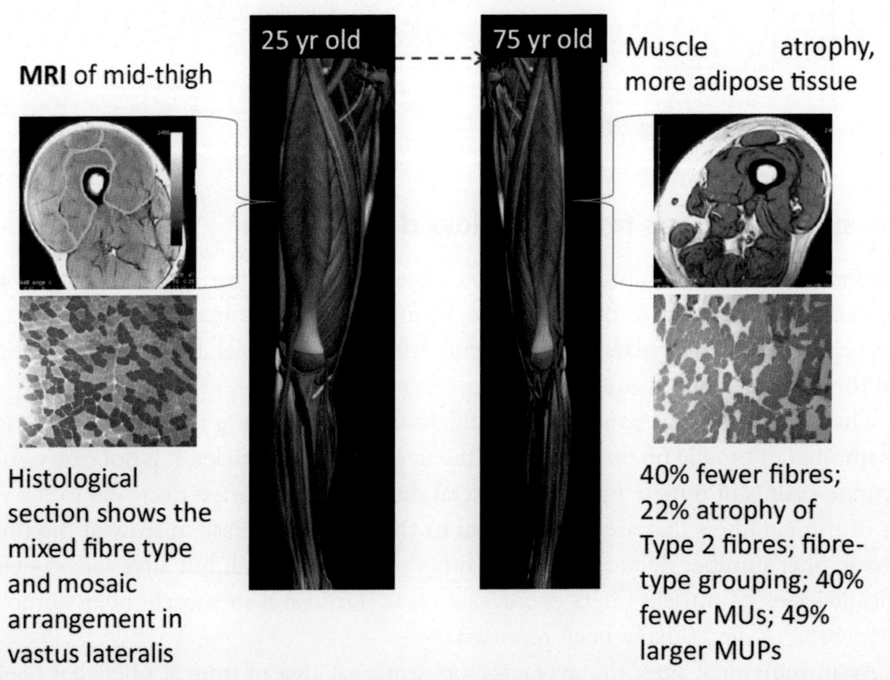

Figure 8.4 Various aspects of atrophic leg muscles in old age. MRI scans show more fat and reduced muscle mass in the older thigh than in a younger one. The histological sections show fiber-type grouping in older muscle. *(From Piasecki et al. (2016) , with permission.)*

Within the muscle fiber, both the dispersion and domains of myonuclei become more irregular with aging. In addition, individual myonuclei in old individuals are more likely to have an irregular shape or be fragmented. Mitochondria are also significantly affected by the aging process. The dynamics of mitochondrial fission and fusion are altered, as is the removal of damaged mitochondria through the process of mitophagy. The accumulation of dysfunctional mitochondria has been suggested to contribute to the atrophy of aging muscle fibers. The number of satellite cells per muscle fiber also decreases with age (see p. 61).

Aging also changes the overall architecture of human muscles. Specifically, the length of muscle fascicles decreases, and the angle of pennation of muscle fibers also decreases. The decrease in fascicle length reflects a loss of sarcomeres and results in a reduced speed and degree of shortening. The decreased angle of pennation, resulting largely from an overall thinning of a muscle (see Fig. 8.4), leads to a reduction of force that can be exerted by the muscle.

The relative amount of interstitial connective tissue, including fat, may actually increase during aging, but this is counterbalanced by a reduction in water content of aging muscle. In very old rats, up to 25% of muscle mass is connective tissue. Increased fat is frequently seen in CT scans of the thighs of elderly humans (see Fig. 8.4), but whether this is a cause or an effect of decreased muscle function has not yet been definitively determined. The role of fatty infiltration of muscle during aging is made even more complex because of associated pathology, especially diabetes, which can in itself lead to increased fat deposition.

In reality, decreases in both muscle fiber numbers and cross-sectional area both appear to play the most important roles in the loss of muscle mass during aging, but the loss of muscle fibers is most important. Mechanisms accounting for both of these are covered later in this chapter.

The role of tendons as factors in the aging of human muscle remains murky. Most structural and mechanical measures of tendons do not change greatly from maturity to old age. One exception is an increased trend in the glycation (binding of sugar molecules) to the amino acids of collagen, with a resulting increase in the distance between adjacent collagen fibers and a possible loss of tensile strength of the tendon. This could lead to strain injuries of a tendon, but that could also be countered by the decreasing amount of force that can be produced by the muscle. Aging tendons also lose water content, a significant factor leading to increased stiffness of old tendons. The microcirculation of a tendon also diminishes during aging, thus impeding the healing of tendon injuries.

Performance measures of aging muscle

Aging affects a variety of performance measures of skeletal muscle. The greater the age, the greater is the variability of performance capability within the group. These measures include force/power, endurance, fatigability, and specific tension, among others. Not all of the factors leading to reduced performance measures are confined to the muscle fibers, and other influences must also be considered to take into account the functional changes in aging muscles. At the gross performance level, changes within the nervous, cardiovascular, and respiratory systems also play significant roles.

The loss of muscle strength with age exceeds that of muscle mass (see Fig. 8.1B), and the loss of muscle power (strength x time) exceeds that of strength. One obvious reason for the loss of strength is loss of mass, because strength is directly proportional to the total physiological cross-sectional area of all of the muscle fibers. Loss of muscle fibers contributes significantly to the loss of strength. Another factor involves the shift in proportions of muscle fiber types. The proportion of Type II fibers is a dominant determinant of the maximum contractile force of a muscle, and the decrease in numbers of Type II fibers, as well as the reduction in their individual cross-sectional area, strongly contributes to the lowering of muscle strength as aging progresses.

An additional factor leading to reduced overall strength of a muscle is reduced specific force of individual muscle fibers and of the muscle as a whole. Most studies suggest a ~20% reduction of specific force in the muscle fibers of old men and somewhat smaller reductions in laboratory rodents. The basis for the reduced specific force has not been completely determined, but reductions in cross-bridge potential and reductions in peak intracellular Ca^{++} flow from the sarcoplasmic reticulum remain strong possibilities.

At the whole muscle level, the presence of denervated muscle fibers and the infiltration of muscles by fat would also account for a reduction in specific force. The speed of contraction is also reduced in old muscle. One factor that would account for this is a reduced rate of myofilament sliding, as well as decreased rates of Ca^{++} uptake into the sarcoplasmic reticulum, but probably more important is the change in the proportion of fast to slow muscle fibers (see below).

Another measure of muscle function that declines with increasing age is **force steadiness**. This is the ability to maintain a given level of contractile activity at a given load. Younger individuals are able to do so with little variability, whereas with advancing age there is increasing variation in contractile force, especially for light loads. The basis for this is not fully understood, but fluctuations in firing of the remaining motor units appear to account for much of the variation in force steadiness.

Fatigue is a complex phenomenon that involves several different levels, from that of the muscle fiber to functions of the CNS. Overall, for other than maximal loads, fatigability may actually be less in healthy older than in younger people for static loads. In contrast, for dynamic loading, fatigue is greater in older individuals.

Endurance performance for most tasks also declines with age, although not to the same extent as does power (see Fig. 8.2). In order to determine what contributes to this decline, one must first ask whether the most important factors are intrinsic or extrinsic to the muscle fibers. Although specific force of Type I muscle fibers does slightly decline with age, available evidence suggests that extrinsic factors may be the most important determinant.

Table 8.1 Measures affecting endurance performance in young and older exercised-trained men.

	Young (28 yrs)	Older (60 yrs)	Age-related change (%)
O_2 consumption (mL/kg/min)	68.2	49.4	−28
Cardiac output (L/min)	27.0	21.7	−20
Stroke volume (mL/beat)	147	132	−10
Heart rate (beats/min)	184	165	−10
A-V O_2 difference (mL/100 mL)	16.7	15.2	−8

Analyses of possible extrinsic factors point to functional declines in both the cardiovascular and respiratory systems as major contributors to the decline in endurance performance. Reduction in the vital capacity of aging lungs has been well documented. Similarly, with increasing age, both the maximum heart rate and stroke volume (amount of blood pumped out with each heartbeat) steadily decline. The major extrinsic influencer of endurance performance is maximal aerobic capacity (V_{O2max}), as measured by the Fick equation (V_{O2max} = maximum cardiac output X maximum difference between arteriovenous O_2 concentration). The collective impact of all these variables is summarized in Table 8.1. Maximum heart rate declines by about 0.7 beats per year (7 beats per decade), and maximum stroke volume is reduced by 10%–20% by old age.

The capillary density around muscle fibers declines somewhat during aging, but the functional significance of this remains controversial. The ability to extract O_2 from the blood in the microcirculation declines somewhat less than 10% over 30 years and is likely related to a decrease in capillary density around the muscle fibers. Overall, central factors (aging changes in the heart and lungs) are more important than peripheral ones (oxygen extraction) as limiters of endurance performance in older individuals.

Neuromuscular interactions in aging muscle

At every point in life, skeletal muscle is dependent upon its innervation for both its maintenance and its function. As aging progresses, interactions between muscle fibers and their nerves play an increasingly important role as determinants of muscle function. These interactions reflect activities within the nervous system that begin with the firing of cortical neurons and continue down to the level of the neuromuscular junction (NMJ). Characteristics of the normal patterns of neuronal firing change with age, but even more important are the connections between nerve fibers and muscle fibers. The most extreme age-related disruption is the result of death of motor neurons. In addition to changes in motor innervation, the sensory innervation of muscles is also affected by the aging process.

Structural and functional changes in the cerebral cortex can affect muscle function through their direct effects upon spinal motor neurons and through their disrupted integration of sensory signals coming from the spinal cord. Structurally, neuronal atrophy, rather than loss of neurons, seems to be a dominant feature of cerebral aging and is evidenced by decreased central command, reduced connectivity and integration, as well as neural noise, which reduces the signal/noise ratio.

The most direct neural influence on aging muscle begins in the spinal cord with the death of alpha motor neurons, resulting in a loss of motor units. In the absence of neuronal death, slower nerve conduction and many structural and functional changes in the NMJ also contribute to aging changes in skeletal muscle.

By definition, neuronal death directly affects the motor unit composition of muscles. Motor neuronal death, which becomes significant after age 60, eliminates a large number of the motor neurons as aging progresses. According to one study, estimated numbers of motor units in the human tibialis anterior muscle declined from 150 in young men, to 91 in men in their 60s, to 59 in men in their 80s. Overall, by age 75, approximately only 60%–70% of the motor units in muscles of the lower limb remain. Motor unit losses up to 85% have been reported in certain muscles of very old individuals. Despite the decline in motor unit number, overall strength is maintained longer than would be expected based on motor unit number alone. This is due to an increase in the overall size of the remaining motor units with increasing age.

Age-related motor neuronal death affects especially the larger alpha motor neurons, while sparing the smaller gamma motor neurons. The immediate consequence of motor neuronal death is denervation of all muscle fibers in the motor unit that was supplied by the dead neuron. Functionally, the smaller the number of motor units in an aging muscle the lesser the degree of fine motor control over the contractions of that muscle. Because many of the noninnervated motor units are those large ones consisting of Type IIx muscle fibers, maximum strength of the muscle declines. The selected functional removal of fast-twitch muscle fibers also tips the balance of the muscle toward slower contractile speed.

Two possible fates await the muscle fibers that are denervated as a result of motor neuronal death. Some of them will become reinnervated by sprouts from the terminal portions of neighboring nerve fibers, with the result that the motor units supplied by these nerve fibers will enlarge (Fig. 8.5). Other muscle fibers will remain denervated and will gradually undergo atrophy and disappear. These denervated muscle fibers will no longer contribute to overall muscle contractile function. The muscle fibers that are reinnervated will take on the characteristics of the other muscle fibers innervated by that same nerve fiber. During aging, capture by terminal sprouting is more effective in slow motor axons, with the result that any denervated Type II muscle fibers that are reinnervated by slow nerve sprouts will take on the characteristics of slow muscle fibers. During the conversion process, many of these muscle fibers will contain a mix of fast and slow myosins. Unfortunately, aging also results in a decreased capacity for sprouting of motor neurons. This increases the potential for more muscle fibers to remain denervated after the motor nerve fiber that originally supplied them died.

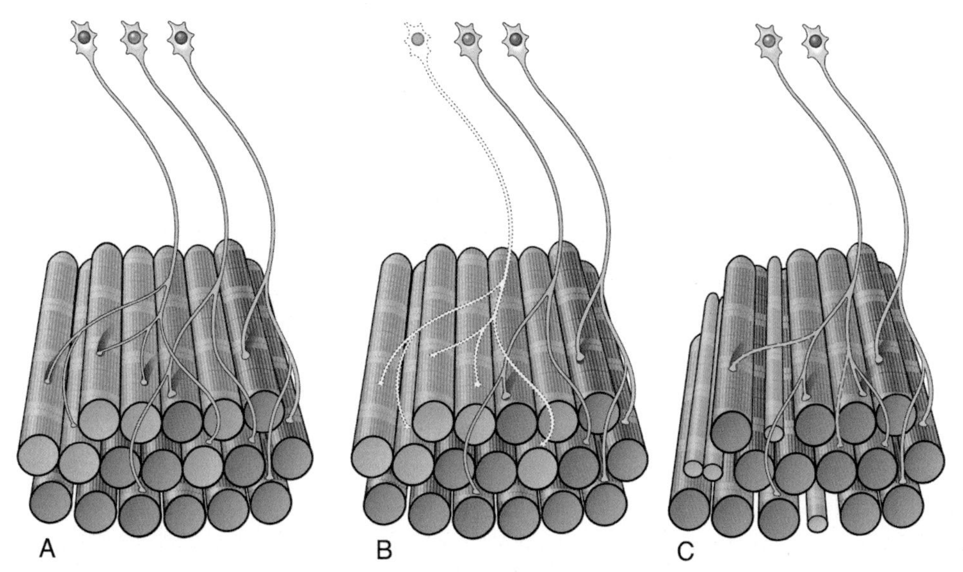

Figure 8.5 Motor unit remodeling in old muscle. (A) Muscle before motor neuron death. (B) When a motor neuron dies (neuron farthest left), the muscle fibers connected to it become denervated. (C) Some of these fibers are reinnervated by sprouts from nearby nerve terminals (bluish neuron); others (thin muscle fibers) remain denervated and undergo atrophy.

The denervation/reinnervation process causes a change in the distribution of muscle fiber types within an old muscle. In a normal young motor unit, the muscle fibers supplied by a single motor axon are intermixed among muscle fibers of many other motor units. This architectural arrangement allows smoother contractions of a muscle, rather than having all muscle fibers in a motor unit contracting in a concentrated, relatively small region of the muscle. During reinnervation of denervated muscle fibers, whether due to aging or disease, the reinnervated muscle fibers are typically juxtaposed to the muscle fibers innervated by the same sprouting nerve fiber. This leads to fiber type grouping (see Fig. 8.5), which is characteristic of old muscle.

A major site where age-related denervation occurs is the NMJ. NMJs are not stable. The overall morphology of NMJs changes (remodeling) at a slow rate at almost all stages of life, but the frequency and extent of remodeling increases considerably during old age. Changes occur in both the presynaptic and postsynaptic parts of the NMJ. Presynaptically, there is an increase in the amount and complexity of terminal branching of the motor nerve fiber, although the number of acetylcholine-containing terminal vesicles decreases. On the postsynaptic side, the overall boundary area of the endplate may actually increase during early old age, but in later old age it decreases, and acetylcholine receptors become more dispersed (Fig. 8.6). The basic pattern is increasing dissociation between the axon terminals and the postjunctional folds, resulting in some NMJs having no morphological contact between the pre- and postjunctional components of the NMJ, making the muscle

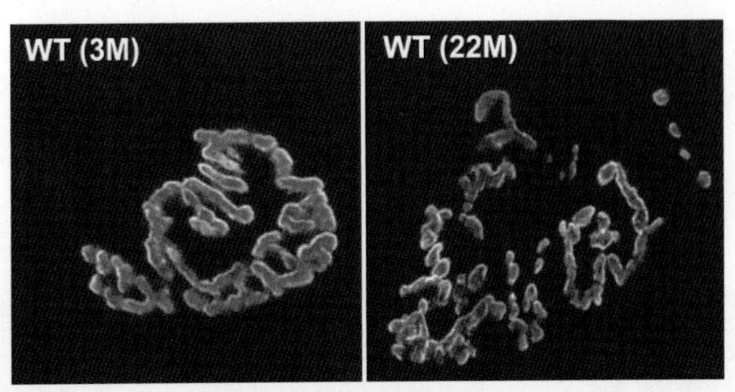

Figure 8.6 Motor endplates in young (3-month) and old (22-month) mice. In old muscle, the motor endplate region undergoes fragmentation and expansion. *(From Larsson et al. (2019), with permission.)*

fiber, in effect, denervated. This, along with less acetylcholine release by the nerve terminals results in less efficient signaling between motor nerve and muscle fiber, although during most of life several times more acetylcholine is released than is necessary to produce an action potential on the surface of the muscle fiber. Paradoxically, during the early stages of aging, there may be an actual increase in presynaptic signaling. The signaling deficit during late aging, as noted above, may not produce demonstrable effects during the early stages of remodeling because of the normal excess production of acetylcholine. Considerable evidence suggests that the loss of motor neuronal support begins at the NMJ and the proceeds up the axon before affecting the neuronal cell body itself.

Aging affects negatively not only motor functions, but also proprioception, which is largely a function of muscle spindles and Golgi tendon organs. One consequence of a decline in proprioception is an increased incidence of falls in the elderly. Indirect evidence suggests a decrease in the number of muscle spindles with aging, and direct evidence has shown fewer intrafusal fibers in muscle spindles, along with structural and functional disturbances in the endings of sensory nerve fibers on the intrafusal muscle fibers. Some intrafusal fibers become denervated, but the spindles affected by denervation may persist in old age as they do after denervation in younger animals. For a basic function, such as standing, decreased proprioceptive input from muscle spindles can be compensated by coactivation of antagonist muscles in order to stabilize the ankle joint.

An important functional aging change is a decrease in conduction velocity in the sensory nerve fibers supplying muscle spindles. In addition to muscle spindles, Golgi tendon organs are negatively affected by the stiffening of tendons in old age. All of these changes in the proprioceptive apparatus of muscle lead toward an increased tendency to lose one's balance and fall. Interestingly, at a standing position, the most important sensory cues are proprioception and vision. The vestibular system plays only a minor role in maintaining balance while standing.

Sarcopenia

"Sarcopenia has been defined as a progressive and generalized skeletal muscle disorder that involves the accelerated loss of muscle mass and function" (The Lancet, 2019).

Ever since the term was coined, workers in the field have struggled to define the condition and to distinguish it from the deterioration in muscle structure and function that is a normal concomitant of aging. Nevertheless, sarcopenia has important consequences. It leads to increased numbers of falls, increased susceptibility to muscle injury and to pathological conditions such as obesity or type II diabetes. Estimates suggest that 10% of adults over 60 years old and up to 50% of people over 80 years old exhibit sarcopenia.

Historically, it has been both difficult to define and to understand factors leading to the pathogenesis of sarcopenia. Characteristic pathological features of sarcopenia are listed in Table 8.2. Unfortunately, no single marker of sarcopenia exists, and there seems to be a continuum between the normal consequences of muscle aging and its pathological extension—sarcopenia—which affects up to 50% of all people over 80 years of age. The practical consequences, however, are significant because individuals affected with sarcopenia are generally frail, have limited mobility, are more susceptible to falls and a number of chronic conditions, and have a higher mortality than nonsarcopenic individuals of the same age cohort.

Sarcopenia is not just a condition seen in old age. Excessive muscle loss is seen in a variety of chronic diseases, for example, kidney disease, tumors, heart failure, malabsorption, and malnutrition. Whether the mechanisms underlying muscle loss in these conditions differ significantly from the excessive muscle loss during aging remain a subject of investigation.

Table 8.2 Hallmarks of sarcopenia.

Decreased muscle mass
Decreased number of muscle fibers
Decreased number of motor units (especially α-motor neurons)
Muscle fiber atrophy (especially Type II muscle fibers)
Decreased speed of contraction (at both whole muscle and muscle fiber level)
Slowing of movement
Intramuscular fibrosis and fat deposition
Decreased protein synthesis in relation to degradation
Decreased mTOR pathway activity
Mitochondrial dysfunction
Increased oxidative stress
Decline in autophagy
Increased expression of FOXO and atrophy genes
Decreased satellite cell numbers and function
Chronic inflammation
Anabolic resistance
Impaired regenerative capacity

Satellite cells and muscle regeneration in old age

The success of skeletal muscle regeneration declines with increasing age and decreases sharply in old age. The decline involves both the rate and robustness of the regenerative response. At a gross level, research studies paint a consistent picture. As one ages, muscles become increasingly susceptible to damage, especially from eccentric contractions. Once an old muscle becomes injured, subtle changes in the inflammatory response result in a slower initiation of both the degenerative and early proliferative phases of the regenerative response. The overall process of regeneration takes longer. Although there are some exceptions, the success of muscle fiber regeneration is also diminished as measured by various aspects of contractile force. In addition, fibrosis is more likely to accompany and reduce the success of the regeneration of old muscle.

Disturbances in many aspects of the regenerative process contribute to this decline. They include age-related changes within the satellite cell population and regenerating muscle cells themselves, as well as changes in the environment surrounding the regenerating muscle cells.

Intrinsic age-related changes in satellite cells and their progeny

Despite their being stem cells, satellite cells are not immune to the effects of aging. Changes in a variety of signaling and metabolic pathways affect both individual satellite cells and their population dynamics. Yet these changes are most commonly the responses of old satellite cells to alterations in their immediate external environment.

Numbers of satellite cells decrease with increasing age, but the most dramatic decreases occur during the last 10% of the normal life span. In very old rats, satellite cell numbers may fall to only 20% of that in young adults. Within a given population, all satellite cells do not appear to be equally affected by aging changes. Some function like those in normal mature individuals, whereas others may be severely affected to the point where they are essentially nonfunctional. In mice, at least, a loss of ability of satellite cells to support muscle regeneration precedes the decline in numbers of satellite cells. This is a reflection of the presence of a subpopulation of senescent satellite cells in old muscle that are no longer able to participate in regeneration.

Satellite cells in old individuals show a number of intrinsic aging changes. Their oxidative capacity is reduced because of the accumulation of increasingly large numbers of dysfunctional mitochondria, which remain in the cell because of a reduced capacity for autophagy (see p. 71)—the means by which damaged mitochondria and macromolecules are removed from a cell. The mitochondrial damage is a reflection of intracellular stress caused by increased exposure to reactive oxygen species. Within the nucleus, epigenetic changes at the level of histones and chromatin structure in old satellite cells result in significant changes in a variety of patterns of gene expression, especially those genes controlling the cell cycle. At a certain point, these changes become irreversible. A cumulative effect of the intrinsic changes in satellite cells is a decline in their proliferative potential (Fig. 8.7). Many other intracellular changes in aging satellite cells are a direct result of changes in their immediate environment. These are discussed in the next section.

Environmental influences on satellite cell aging

The importance of the environment on the aging of satellite cells and on muscle regeneration in old animals was demonstrated by exposing regenerating old muscles to young environments. This was first demonstrated by grafting entire muscles from old inbred rats into young individuals, along with grafting young muscles into old hosts (Box 8.1; Fig. 8.8A). Old muscles grafted into young rats regenerated nearly as well as young muscle autografts, whereas young muscles grafted into old rats regenerated no better than did old muscle autografts (Fig. 8.8B). This experiment showed that the environment in which a muscle regenerated played a very important role in the success of the regenerative process.

A similar approach involved making parabiotic[1] pairs of old and young mice, and subjecting muscle in members of the pair to damage. In this case, the damaged muscle in the old parabiotic mouse regenerated somewhat better than would be expected, and in the younger mouse, muscle regeneration was somewhat less successful than that seen in a normal young mouse.

In the case of the cross-age muscle transplantation experiment, innervation was a significant environmental factor, because in this model ultimate success of the muscle grafts depended upon the success of reinnervation of the muscle grafts. Nevertheless, other environmental factors also played a role. In the parabiotic experiment, the dominant environmental factor was the blood to which the regenerating muscles were exposed. It showed that something present in the circulating blood of both old and young mice was able to affect the success of the regenerative process. Presumably, the blood of young mice contains regeneration-promoting factors (e.g., oxytocin or growth differentiation factor −11[GDF-11]), whereas blood of old mice contains factors (e.g., Wnt-3 and TGF-β) that actually inhibit regeneration.

Since these in vivo experiments were conducted, the more recently developed ability to isolate specific populations of satellite cells by various cell-sorting techniques has

[1] In a parabiosis experiment, a patch of skin along the side of two animals is removed, and the edges of the skin of the two animals are sutured together so that the animals are locked in a side-by-side position. Over time blood vessels regenerate into the area where the two animals have been fused, and they develop a common circulation. When animals of a different age are parabiosed, the young animal is exposed to some of the blood of the older animal, and the older animal is exposed to some of the younger blood.

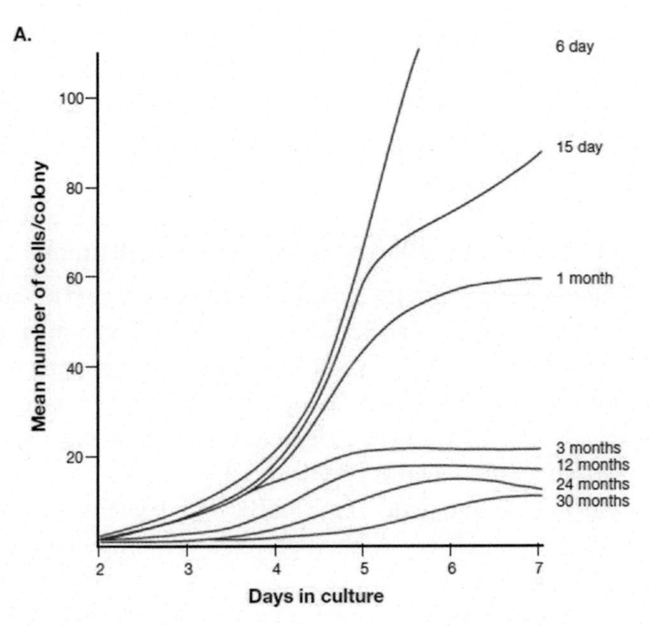

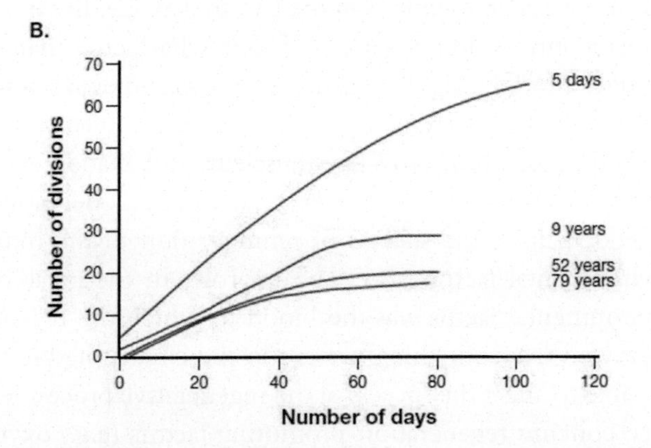

Figure 8.7 Changes in the proliferative potential of satellite cells with age. (A) Mean numbers of satellite cells per colony over days of culture derived from single satellite cells taken from rats of different ages. (B) Numbers of divisions of cultured satellite cells over time from satellite cells taken from human donors of different ages. *(From Carlson (2007), with permission.)*

allowed more sophisticated molecular analysis of aging changes in satellite cells and the role of the environment in producing these changes.

A significant age-related change involves the Delta-Notch interaction (see p. 61). In this case, the environmental factor is the muscle fiber itself. As muscle fibers age, the Delta ligand on the sarcolemma of muscle fibers strongly diminishes. One of the inhibitory actions of Wnt-3 in old muscle is a reduction in Notch signaling (Fig. 8.9). The reduction in Notch reduces their quiescence and promotes the activation and differentiation of satellite cells. Ultimately, this reduces the number of satellite cells available to participate in regeneration.

BOX 8.1 Muscle transplantation

Free muscle transplantation consists of completely removing a muscle from its bed by transecting the proximal and distal tendons, along with severing all blood vessels and nerves. The completely removed muscle can then be grafted back into its original location (orthotopic graft) or into a different location (heterotopic graft). In rats and mice, this procedure results in the degeneration of over 98% of the muscle fibers from ischemic necrosis. A thin outer rim of muscle fibers survives through the diffusion of oxygen in the interstitial fluids. By soaking a removed muscle in a myotoxic agent (e.g., Marcaine [bupivacaine]), one can obtain the death of 99.9% of the original muscle fibers in the graft. Following revascularization from the centripetal ingrowth of blood vessels from the periphery, a gradient of muscle fiber degeneration/regeneration is established (Fig. 6.5). This model is useful for studies on contractile properties of regenerating muscle because of the lack of contamination of undamaged muscle fibers along with regenerating fibers.

In a pure free muscle graft, axons grow out from the stump of the transected nerve without the benefit of preformed pathways. Such a regenerate ultimately recovers about 30%–40% of the tetanic force of the original muscle. On the other hand, if the motor nerve branch into the muscle is not severed during the grafting process (nerve-intact graft), the grafted muscle in a young animal ultimately returns to over 95% of its original tetanic force because the motor nerve fibers, even if they degenerate and later regenerate, are able to utilize preexisting pathways to the muscle fibers. This ensures a more robust distribution of nerve fibers to the vast majority of the regenerating muscle fibers.

Other age-related environmental effects associated with muscle fibers themselves (as well as associated fibroadipogenic progenitor cells [FAPs]) are an increase in muscle fiber-produced FGF-2 and TGF-β, both of which interfere with muscle regeneration and promote fibrosis. These growth factors, working through increased activity of their respective signal transduction pathways (ERK and SMAD), act to push satellite cells from their quiescent state, but interfere with their proliferation. FGF-2 inhibits the intracellular protein Sprouty-1, which is necessary for self-renewal of quiescent satellite cells during regeneration. As a result, dividing satellite cells are pushed to produce two differentiating myoblasts rather than engaging in an asymmetrical division that would result in one myoblast and one stem cell that would help maintain the satellite cell population. Ultimately such a lack of replenishment diminishes the pool of satellite cells in aging muscle.

The serum environment surrounding aging satellite cells, especially after muscle damage, is awash with a cocktail of cytokines and growth factors. Some of these are products of the chronic inflammatory cells, which increase in numbers during the aging process. Elevated levels of TNF-α, associated with M1 macrophage activity, stimulate the activation and proliferation of satellite cells. This, however, is counteracted by a decrease in the

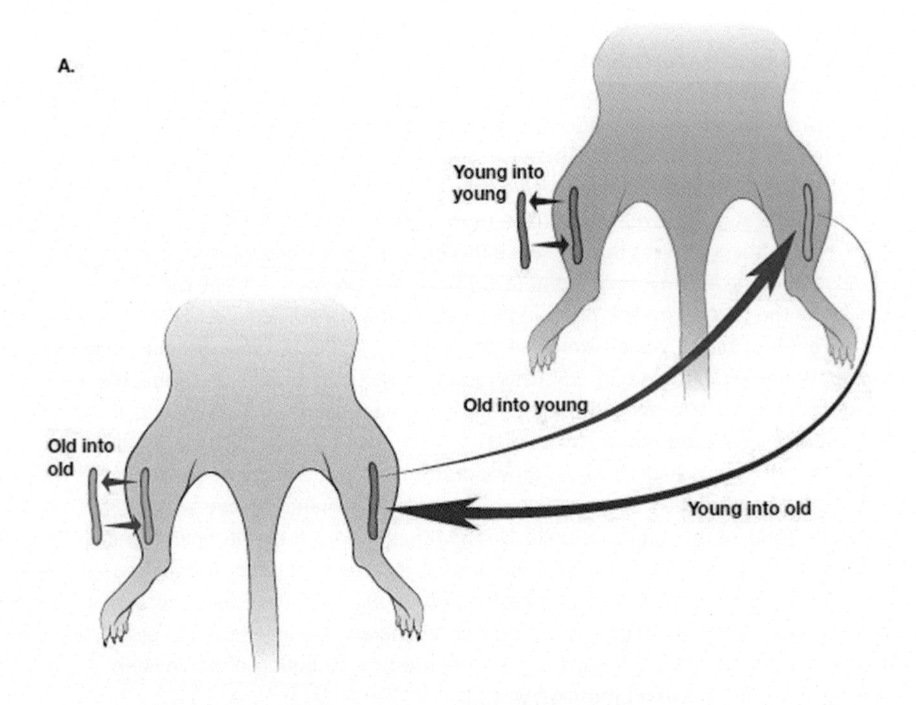

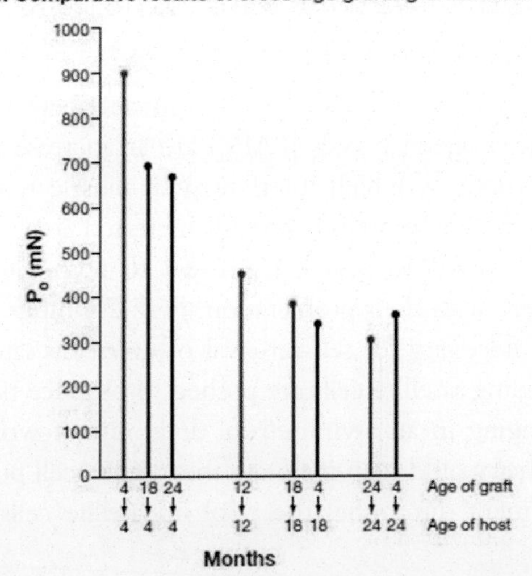

Figure 8.8 (A) Scheme of cross-age muscle transplantation between young and old rats. (B) Comparison of functional results of cross-age grafts (*black*) of the rat extensor digitorum longus muscle versus same-age grafts (*red*). The maximum tetanic force (P_o) of same-age grafts steadily declines with age. The P_o of cross-age grafts is closer to that of same-age grafts of the recipient than that of the donor. *(From Carlson (2007), with permission.)*

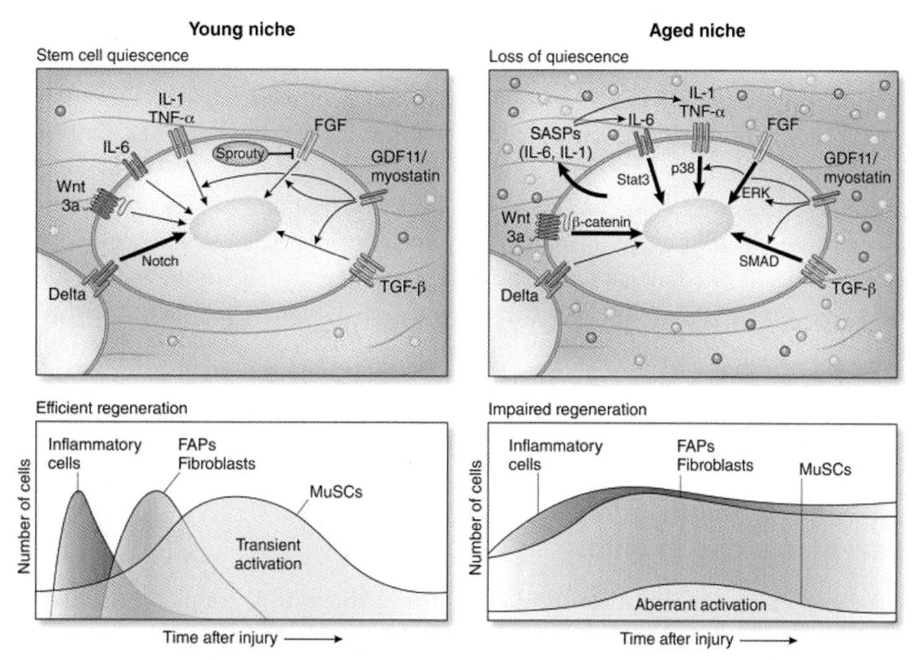

Figure 8.9 Extrinsic and intrinsic regulators of satellite cell function after tissue damage in young and old muscle. Much of the age-related impairment of repair is due to a continued increased level of inflammatory cells and a corresponding increase in activity of FAPS (fibroadipoprogenitor cells). *FGF*, fibroblast growth factor; *GDF*, growth differentiation factor; *IL*, interleukin, *MuSCs*, muscle satellite cells; *TNF-α*, tumor necrosis factor. *(From Blau et al. (2015), with permission.)*

promyogenic cytokine IL-1 and an increase in the antimyogenic cytokine IL-6 and growth factor Wnt-3. Wnt-3 additionally promotes fibrosis by stimulating muscle stem cells and FAPs to contribute to fibrosis rather than muscle fiber regeneration. Compounding these effects are reduced serum levels of the regeneration-promoting hormone oxytocin in old individuals.

The net result of the internal responses of satellite cells to external influences is an age-regulated distribution of key intracellular signaling pathways (p38α/β MAPK) so that elevated levels of these pathways are symmetrically distributed to each daughter cell of a satellite cell division. (In young satellite cells, the distribution is asymmetrical so that the daughter cell that has high activity of these pathways goes on to differentiate, whereas the other daughter cell with less activity returns to quiescent satellite cell status.) This results in two significant consequences. First, both daughter cells go on to differentiate, thus leaving no quiescent satellite cell after the division. Ultimately, this leads to exhaustion of the satellite cell population. Second, the same internal environment activates a cell cycle inhibitor ($p16^{INK4A}$) that propels the cell into senescence.

Even physical changes in the immediate extracellular environment of aging satellite cells alter their response to injury. The extracellular matrix surrounding muscle fibers becomes stiffer with age. Increased cross-linking of collagen fibers accounts for much of the stiffness. Such a mechanical environment promotes differentiation, rather than the proliferation of satellite cells.

Intrinsic changes to aging satellite cells and corresponding changes in their extracellular environment combine to diminish the response of a muscle's satellite cell population to muscle injury. The result is a reduction in regenerative capacity of the muscle. Changes in individual satellite cells ultimately affect population dynamics, thus affecting the regenerative ability of an entire muscle. Changes in the patterns of proliferation and differentiation of satellite cells affect principally early stages of regeneration, whereas increased fibrosis and a decreased success of reinnervation of regenerating muscle fibers negatively affect later stages of regeneration.

Effects of exercise in old age

The steadily increasing numbers of older adults in the world, coupled with the documented declines in neuromuscular function during the aging process, have focused attention on ways to reduce or reverse these declines. Other than for reasons of individual well-being, improving neuromuscular function of the elderly would have a significant economic impact because it would decrease the incidence of injuries due to falls and also increase the level of functional independence of a significant number of old people. By far, the most effective means of accomplishing this goal is through sensible exercise programs involving both endurance and resistance training.

The major muscular issue that is addressed by exercise is loss of strength, but endurance, balance, and reaction speed can also be significantly improved. Exercise affects essentially every system in the body, especially the cardiovascular and respiratory systems. At the metabolic level, exercise plays a significant role in counteracting insulin resistance, the primary pathology underlying type 2 diabetes.

Overall, exercise in the elderly produces similar effects to those seen at any age, but typically the response in the elderly is somewhat blunted compared with that seen in younger ages. Individual responses to exercise programs in humans vary considerably. For a given exercise protocol, some subjects (superresponders) show pronounced improvement, whereas others (nonresponders) fail to change. Genomic studies have shown significant differences in gene expression between superresponders and nonresponders. There appears to be no age at which exercise fails to produce measurable improvements in muscle performance. This has been demonstrated in studies on nonagenerians in group living facilities.

It is important to recognize the effect of exercise on the course of decline in muscle function that occurs during the aging process. If one looks at the typical slope of performance decline in a sedentary population, a well-designed exercise program can significantly raise one's level of performance, but once that has been attained, the slope of decline continues at the same rate as before, but the level is now higher (Fig. 8.10).

Both endurance training and resistance training produce substantial improvement of muscle function in the elderly, and a well-balanced exercise program incorporates both. Much research has gone into the preparation of guidelines for effective exercise for the elderly. Recommendations from the American College of Sports Medicine and the National Strength and Conditioning Association are summarized in Table 8.3.

Many of the positive effects of exercise in the elderly are similar to those seen in younger individuals, but individual variability tends to be greater. A significant factor leading toward variability involves conditions secondary to aging, for example, associated diseases, chronic inflammation, motion-limiting pain, or even inactivity. For some individuals, exercise programs can bring their muscular function back to levels of sedentary populations 20 years younger in age. For others, exercise only slows down the slope of muscular functional decline.

One of the most prominent results of endurance training in the elderly is either maintenance or improvement of mitochondrial biogenesis and oxidative activity, but the age-related decrease in mitochondrial oxidative capacity is not fully reversible. At the metabolic level, some of the exercise-mediated changes are accomplished through their effects on PGC-1α (see p. 88). Among the beneficial effects, especially with respect to type 2 diabetes, is an increase in glucose uptake (insulin sensitivity) after even an hour of walking. At a more systemic level, aerobic exercise increases V_{O2max} and peak working capacity, among other elements of cardiopulmonary function. The research literature presents a variety of views on the effect of endurance exercise on local capillarization of muscle fibers and oxygen transfer between the microcirculation and muscle fibers.

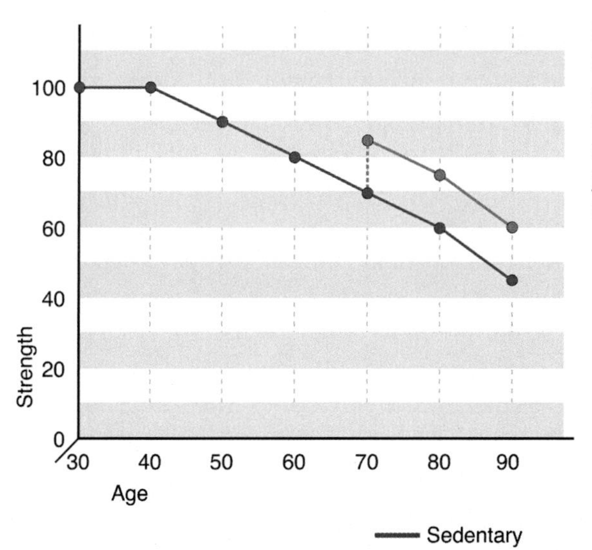

Figure 8.10 Graph showing an increase in strength following an exercise program in a 70 year old. However, the rate of decline in strength does not change with increasing age. Strength just remains at a higher level than in sedentary subjects.

Table 8.3 Recommendations[a] for parameters of exercise programs for older adults.

Function	Details
Endurance exercise	
Frequency	Bouts of 30–60 min leading to 150–300 min per week
Intensity	On a scale of 1–10, between moderate (5–6) and vigorous (7–8)
Duration	For moderate levels, at least 30 min per day; for vigorous, at least 20 min per bout
Resistance exercise	
Frequency	At least two times per week
Intensity	On a scale of 1–10, between moderate (5–6) and vigorous (7–8)
Type	Progressive weight training or weight bearing 8–10 exercises involving the major muscle groups, 8–12 reps each
Flexibility exercise	
Frequency	At least two times per week
Intensity	On a scale of 1–10, moderate (5–6)
Type	Many forms of stretch

[a]Adapted from Chodzko-Zajko et al. (2009).

Effective resistance training produces a number of well-defined changes in neuromuscular function. At any age, it results in significant improvements in overall muscle strength. Not all of the improvement can be attributed to muscular changes. Increase in muscle strength and power precedes a morphological response. Any improvement during the first weeks of a resistance training program is most likely due to neural adaptation and is seen before demonstrable muscular changes are evident. One of the bases for the long-term improvement in muscle strength through resistance exercise is hypertrophy of all types of muscle fibers. The balance of muscle fiber types shifts from Type IIX to IIA. There is little evidence for hyperplasia of muscle fibers.

Those elderly persons who respond best to resistance training show a robust increase in metabolic pathways leading to protein synthesis, especially activation of the mTOR complex. This is facilitated by increases in rRNA synthesis, but at a much lower level than what occurs in young individuals. Age-related differences in protein breakdown, however, are much less pronounced.

Of importance for older adults who have embarked upon an exercise program is the follow-up. In the absence of continuing training, increased dynamic strength is maintained at some level for several months, but return to control levels is inevitable. Most trainers recommend that older people continue with at least minimum maintenance exercise in order not to lose the benefits gained from a previous exercise program. These benefits are not confined to skeletal muscle. They affect most systems in the body at both the cellular (Fig. 8.11) and the organ level. One of the most difficult problems in the public health domain is finding means of stimulating not only the elderly, but the entire population, to abandon a sedentary lifestyle in favor of a more active one.

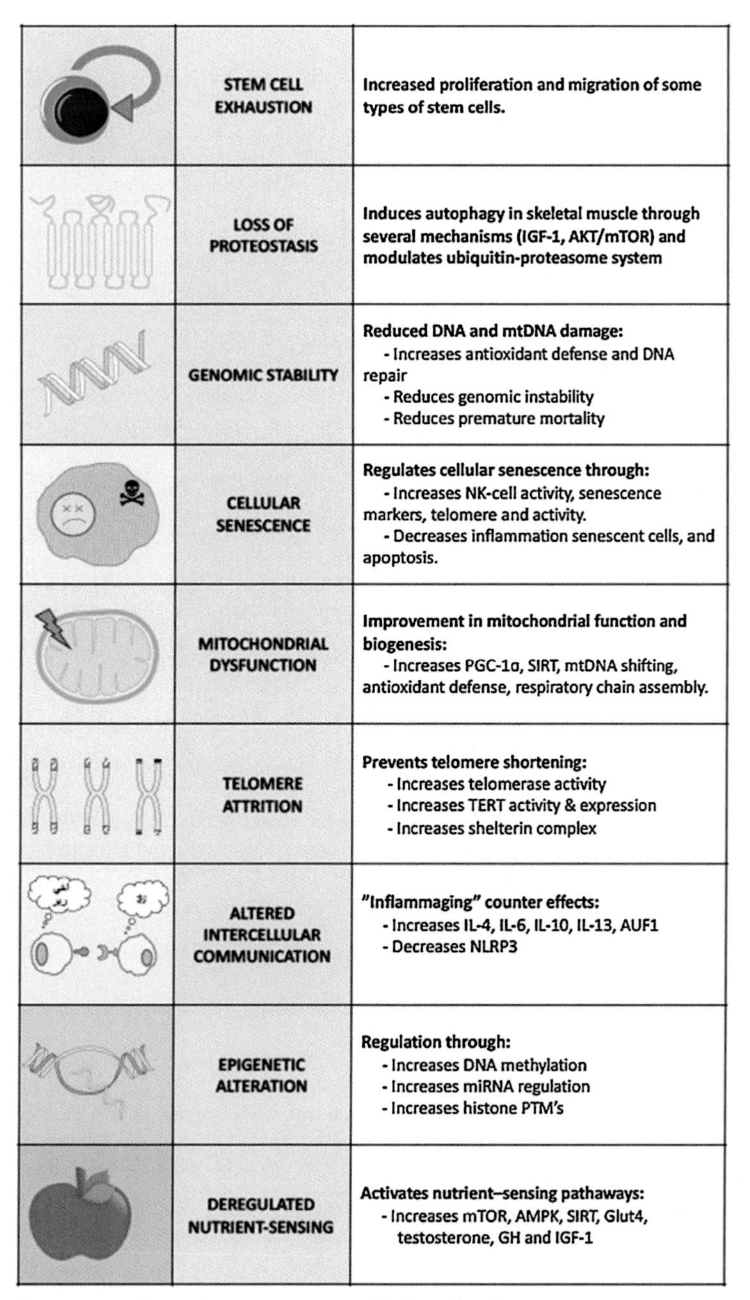

Figure 8.11 The impact of exercise on a variety of hallmarks of aging. *(From Navas-Enamorado et al. (2017), with permission.)*

References

Cartee GD, Hepple RT, Bamman MM, Zierath JR. Exercise promotes healthy aging of skeletal muscle. Cell Metabol 2016;23:1034–47.

Chodzko-Zajko WJ, Proctor DN, Singh MAF, Minson CT, Nigg CR, Salem GJ, Skinner JS. Exercise and physical activity for older adults. Med Sci Sports Med 2009;41. 1510-153.

Close GL, Kayani A, Vasilaki A, McArdle A. Skeletal muscle damage with exercise and aging. Sports Med 2005;35:413–27.

Cruz-Jantoft A, Sayer AA. Sarcopenia. Lancet 2019;393:2636–46.

Deschenes MR. Motor Unit and neuromuscular remodeling with aging. Curr Aging Sci 2011;4:209–20.

Domingues-Faria C, Vasson M-P, Goncalves-Mendes N, Boirie Y, Walrand S. Skeletal muscle regeneration and impact of aging and nutrition. Aging Res Revs 2016;26:22–36.

Fragala MS, Cadore EL, Dorgo S, Izquierdo M, Kraemer WJ, Peterson MD, Ryan ED. J Strength Condit Res 2019;33:2019–52.

Garatachea N, Pareja-Galeano H, Sanchis-Gomar F, Santos-Lozano A, Fiuza-Luces C, Moran M, Emanuele E, Joyner MJ, Lucia A. Exercise attenuates the major hallmarks of aging. Rejuvenation Res 2015;18:57–89.

Harridge SDR, Lazarus NR. Physical activity, aging and physiological function. Physiology 2016;32: 152–61.

Hepple RT, Rice CL. Innervation and neuromuscular control in ageing skeletal muscle. J Physiol 2016;594: 1965–78.

Hikida RS. Aging changes in satellite cells and their functions. Curr Aging Sci 2011;4:279–2907.

Hunter SK, Pereira HM, Keenan KG. The aging neuromuscular junction and motor performance. J Appl Physiol 2016;121:982–95.

Janssen I, Heymsfield SB, Wang Z, Ross R. Skeletal muscle mass and distribution in 468 men and women aged 18-88 yr. J Appl Physiol 2000;89:81–8.

Larsson L, Degens H, Li M, Salviati L, Lee Y, Thompson W, Kirkland JL, Sandri M. Sarcopenia: aging-related loss of muscle mass and function. Physiol Rev 2019;99:427–511.

Lavin KM, Roberts BM, Fry CS, Moro T, Rasmussen BB, Bamman MM. The importance of resistance exercise training to combat neuromuscular aging. Physiology 2019;34:112–22.

Roman MA, Rossiter HB, Casaburi R. Exercise, aging and the lung. Eur Respir J 2016;48:1471–86.

Rygiel KA, Picard M, Turnbull DM. The ageing neuromuscular system and sarcopenia: a mitochondrial perspective. J Physiol 2016;594:4499–512.

Siu PM. Muscle apoptotic response to denervation, disuse, and aging. Med Sci Sports Exerc 2009;41: 1876–86.

Socha MJ, Segal SS. Microvascular mechanisms limiting skeletal muscle blood flow with advancing age. J Appl Physiol 2018;125:1851–9.

Svensson RB, Heinemeier KM, Couppe C, Kjaer M, Magnusson SP. Effect of aging and exercise on the tendon. J Appl Physiol 2016;121:1353–62.

Tanaka H, Seals DR. Endurance exercise performance in Masters athletes: age-associated changes and underlying physiological mechanisms. J Physiol 2008;586:55–63.

Tieland M, Trouwborst I, Clark BC. Skeletal muscle performance and aging. J Cach Sarcopen Muscle 2018; 9:3–19.

Venturelli M, Reggiani C, Richardson RS, Schena F. Exerc Sport Sci Rev 2018;46:188–94.

Willadt S, Nash M, Slater C. Age-related changes in the structure and function of the mammalian neuromuscular junctions. Ann N Y Acad Sci 2017;1412:41–53.

Glossary

α-Actinin: A protein found in the Z-disk that cross-links terminal parts of actin molecules

A band: The optical band in a sarcomere that is dominated by thick filaments composed of myosin

Acetylcholine receptor: The receptor on the postsynaptic side of the neuromuscular junction that binds acetylcholine

Acetylcholine: The neurotransmitter that functions at the neuromuscular junction

Acetylcholinesterase: An enzyme situated in the basal lamina of the neuromuscular junction that breaks down excess acetylcholine

Action potential: A self-propagating electrical charge that sweeps along the plasma membrane of a neuron or muscle fiber

Adipocyte: A fat cell

Agrin: A molecule that is heavily involved in the clustering of acetylcholine receptors during the development of a neuromuscular junction

Akt/mTOR pathway: A major metabolic pathway leading to protein synthesis in a muscle fiber

Anabolic steroid: A steroid molecule, such as testosterone, that promotes anabolic (synthetic) metabolic processes

Anatomical cross-sectional area: The cross-sectional area of a muscle, as measured perpendicularly to the long axis of the muscle

Apicobasal division: A division of satellite cells in which the mitotic spindle is oriented perpendicularly to the muscle fiber and in which the daughter cells have different developmental fates

Aponeurosis: A broad attachment of a muscle into a bone or fascia

Arteriole: Small-bore arteries that play an important role in regulating blood flow to muscles

ATP (adenosine triphosphate): A molecule that provides energy by cleaving a high energy phosphate bond

Atrogene: One of a family of genes that produces molecules promoting muscle fiber atrophy

Atrophy: The reduction in size of a muscle due to disuse or disease

Autophagy: The breakdown of intracellular particles within a membrane

Basal lamina: A thin layer of extracellular matrix that surrounds a muscle fiber or underlies an epithelial layer

Bipennate: A muscle with muscle fibers extending from a central shaft much like a feather

Blastema (muscle and limb): A group of developmentally young cells that will give rise to a mature structure, for example, a muscle in the embryo or a regenerated limb

Blastocyst: An early human embryo with a fluid-filled central cavity

Blastomere: A cell in an early cleaving embryo

BMP (bone morphogenetic factor): A growth factor in the TGF-β family that is heavily involved in many developmental processes

Bouton: The expanded terminal part of an axon

Cadherin: A cell adhesion molecule that depends upon Ca^{++} for its function

Calmodulin: A muscle fiber protein that binds Ca^{++}

Calpain: A Ca^{++} protease

Calpastatin: A calpain inhibitor

Calsequestrin: A Ca^{++}-binding protein located in the sarcoplasmic reticulum

Capillary: The smallest-bore blood vessels within the circulatory system

Caspase: An enzyme involved in breakdown of muscle proteins

Caveolae: Tiny membrane-bound invaginations of the plasma membrane that play multiple roles in cell physiology

Cell cycle: The life cycle of a cell in four main stages: G_1 (growth), S (DNA synthesis), G_2 (preparation for mitosis), and M (mitosis)

Collagen: A family of important fibrous proteins that make up much of the connective tissue of muscle
Common muscle mass: The group of cells in a limb bud that will later break up to form individual muscles
Concentric contraction: A muscle contraction in which a muscle shortens as it contracts
Costamere: A component that connects sarcomeric proteins to the cell membrane
CPK (creatine phosphokinase): An energy-producing molecule in muscle that leaks into blood plasma after muscle fiber damage
Cross-bridge: Connections between thick and thin filaments that form the basis of muscle fiber contraction
Cytokine: Molecules secreted mainly from inflammatory cells that guide many healing and developmental processes
Dedifferentiation: The production of myogenic cells through the fragmentation of existing multinucleated muscle fibers
Delta: A ligand that binds to notch and is involved in cell differentiation decisions
Denervation: Severing the nerve supply to a muscle
Dermomyotome: A subdivision of a somite that will give rise to muscle and dermis
Desmin: An intermediate filament in muscle fibers
Detraining: Cessation of an exercise program and its effect on muscle
Differentiation: The specialization of a cell during a developmental process
Disuse atrophy: The reduction in size of a muscle that is not being used
Dystrophin: A large protein that connects sarcomeric proteins to the sarcolemma
Dystrophin–glycoprotein complex: A complex of proteins that connects sarcomeric proteins to the extracellular matrix surrounding a muscle fiber
Eccentric contraction: A muscle contraction in which the muscle lengthens as it contracts
Ectoderm: The outer germ layer of an early embryo
Edema: Early swelling after an injury
Electron transport chain: A well-ordered group of molecules on the cristae of mitochondria that are involved in oxidative metabolism
Endocrine: A hormone that is secreted into the blood
Endoderm: The innermost germ layer of an early embryo
Endomysium: The thin layer of connective tissue immediately surrounding a muscle fiber
Enthesis: The connection between a tendon and bone
Epaxial muscle: Trunk muscle located along the back
Epigenetic: Influences beyond the structure of DNA that can influence the products of gene expression
Epimorphic regeneration: A regenerative process involving a regeneration blastema as the source of cells of the regenerate
Epimysium: The outermost layer of connective tissue covering a muscle
Excitation-contraction coupling: The physiological connection between an action potential and Ca^{++} release in a muscle fiber
Excursion: The topographical extent of a muscle contraction
Extracellular matrix: The connective tissue outside a muscle fiber
Extrafusal muscle fiber: A muscle fiber that is not part of a muscle spindle
FAP (fibroadipogenic progenitor cell): A cellular precursor of adipose or fibrous connective tissue
Fascia: A layer of connective tissue that separates other tissues
Fascicle: A small bundle of muscle fibers
FGF (fibroblast growth factor): Member of a large family of growth factors, often promoting cell proliferation
Fibrillation: Tiny involuntary contractions of denervated muscles
Fibroblast: The main cell in connective tissue, which produces collagen and other materials of the ECM
Fibrocartilage: A variety of cartilage in which the matrix is heavily infiltrated with bundles of collagen fibers
Fibrosis: A healing process dominated by the formation of masses of dense fibrous connective tissue

Flaccidity: Lack of contractile activity of a muscle, often after denervation
Force steadiness: The ability to maintain a constant level of contraction while bearing a light load
FoxO: An important molecule leading muscle fiber metabolism into an atrophic process
Fusion: In physiology, the generation of a constant muscle contraction by combining the effects of many successive twitches. In development, the incorporation of a satellite cell or myoblast into a multinucleated muscle fiber
Gastrulation: An early embryonic process leading to the formation of the three germ layers
Golgi tendon organ: A proprioceptive structure located in tendons
Glycation: The addition of carbohydrate side chains onto proteins
Glycogen: The main storage form of carbohydrates in muscle
Glycogenosis: A form of disease involving the improper metabolism of glycogen
Gower's sign: In Duchenne muscular dystrophy, crawling the hands up the legs in order to stand up
Growth factor: A secreted molecule that has a stimulatory or inhibitory effect on a developmental process
Growth hormone: A pituitary hormone that stimulates growth
H band: A band within a sarcomere containing thick, but not thin filaments
Heavy chain: The backbone of a myosin molecule
Histogenesis: The process of forming tissues from precursor cells
Hormone: A secreted peptide or steroid molecule that exerts a distant effect on other tissues
Hypaxial muscle: Muscle of the ventral body wall
Hyperplasia: An increase in the number of cells
Hypertrophy: An increase in the size of a structure, e.g., a muscle fiber
I band: A band in a sarcomere containing thin, but not thick filaments
Id: A transcription factor that inhibits myogenesis
IGF (insulin-like growth factor): A major growth factor that stimulates muscle metabolism
Inflammatory myopathy: A type of muscle disease characterized by large deposits of inflammatory cells around muscle fibers
Integrin: Member of a family of molecules embedded in the plasma membrane and that connects a cell to the surrounding ECM
Integrin–vinculin–talin complex: A complex of molecules functionally associated with integrins
Intrafusal muscle fiber: A muscle fiber located within the capsule of a muscle spindle
Ion channel: A molecular complex that permits the passage of ions through a membrane
Ischemia: A reduction in the blood supply to a structure
Ischemic necrosis: The death of cells due to a prolonged insufficiency of their blood supply
Isoform: A form of protein that is specific for a particular stage in development
Isometric contraction: A contraction in which the muscle neither shortens nor lengthens
Junctional fold: A membrane fold, rich in acetylcholine receptors, in the postsynaptic side of a neuromuscular junction
Knockout: Experimental inactivation of a specific gene
Laminin: A major component of basal laminae
Lateral plate mesoderm: Part of the mesodermal layer in an early embryo that gives rise to a few muscles in the neck
Ligand: A molecule that binds to a receptor
Light chain: A variety of myosin that attaches to heavy chain myosins in a thick filament
LINC (linker of nucleoskeleton and cytoskeleton): A protein that connects components of the nucleus and cytoplasm
Lower motor neuron lesion: Something causing damage or death to motor neurons originating in the spinal cord
Lysosome: Cytoplasmic structure containing many lytic enzymes
M line: A thin line at the very center of a sarcomere

Macrophage: An inflammatory cell that is highly phagocytic and that produces many cytokines
Mast cell: An early responding inflammatory cell that produces histamine
Mesoderm: The middle germ layer of an early embryo
Microcirculation: The complex of small blood vessels that supply a muscle fiber or group of muscle fibers
MicroRNA (miRNA): A small noncoding RNA that influences developmental events at the posttranscriptional level
Mitochondrial biogenesis: Increasing the population of mitochondria through a variety of means
Mitochondrion: An intracellular organelle that plays a key role in oxidative metabolism
Mitophagy: A means of removing damaged mitochondria through engulfment
Morphogenesis: The developmental process involved in taking form of a structure
Motor endplate: The muscular side of a neuromuscular junction
Motor nerve: A nerve carrying contractile impulses to a muscle fiber
Motor unit: A single motor neuron and all of the muscle fibers that it innervates
mTOR (target of rapamycin): A key regulator of the protein—synthetic pathway in muscle
Muscle fiber: A single multinucleated muscle cell
Muscle spindle: A proprioceptive structure located within a muscle
Muscular dystrophy: A generic term for a genetically based muscle disease affecting the function of a variety of key muscle proteins
Myasthenia gravis: A disease involving malfunction of the neuromuscular junction
Myasthenic disorder: A generic term for a disease involving the neuromuscular junction
Myoblast: A precursor cell of muscle
Myocyte: A myoblast that is committed to forming muscle
Myofibril: a bundle of sarcomeric proteins within a muscle fiber
Myofibroblast: A cell appearing during wound healing or regeneration that can cause fibrosis
Myogenesis: The process of muscle development
Myogenic progenitor cell: A developmentally early stage of a cell that will go on to form muscle
Myogenic regulatory factor: A member of a group of transcription factors that guide cells to forming muscle
Myoglobin: An oxygen-binding molecule within a muscle fiber
Myomaker: A protein involved in satellite cell fusion
Myomerger: A protein involved in satellite cell fusion
Myonuclear domain: The amount of sarcoplasm that is served by a single myonucleus
Myonucleus: A nucleus within a muscle fiber
Myopathy: A disease of muscle
Myosin: A major contractile protein
Myositis: Inflammation of muscle
Myostatin: A natural inhibitor of muscle growth
Myotendinous junction: The connection between muscle fibers and tendon
Myotome: That part of a somite that gives rise to muscle precursor cells
Myotonia: A pathological condition of muscle characterized by delayed relaxation
Myotube: An intermediate stage in muscle fiber development
Nebulin: A protein involved in maintaining sarcomeric structure
Necrosis: The death of cells or tissues
Neural adaptation: An increase in muscle function occurring before intramuscular adaptations have time to occur
Neuromuscular junction: A form of synapse between a nerve terminal and a muscle fiber
Neutrophil (polymorphonuclear leukocyte, PMN): The dominant type of acute inflammatory cell
Noncoding RNA: See microRNA
Notch: The receptor of delta

Nuclear bag fiber: A muscle fiber type, characterized by a tight bunch of nuclei, found in muscle spindles
Nuclear chain fiber: A muscle fiber type, characterized by a line of nuclei, found in muscle spindles
Pacinian corpuscle: A deep pressure receptor
Parabiosis: The surgical fusion of two animals, leading to a common blood supply
Paracrine: A local hormonal effect
Paraxial mesoderm: The early region of mesoderm that gives rise to somites
Pax: A member of a family of transcription factors important in satellite quiescence
Perimysium: The connective tissue that surrounds muscle fascicles
Phagophore: A membrane-bounded intracellular structure involved in removing and digesting damaged mitochondria and proteins
Phosphorylase: An enzyme involved in the breakdown of glycogen
Physiological cross-sectional area: The cross-section of a muscle in a plane perpendicular to the long axis of the muscle fibers
Placental insufficiency: A deficiency of placental function leading to poor nutrition of the fetus
Planar division: A division of satellite cells in which both progeny share the same fate
P_o (maximum tetanic force): The greatest force produced by a maximally stimulated muscle
Polyubiquitin chain: A string of ubiquitin molecules attached to a damaged protein
Primary muscle fiber: The first muscle fibers to form in an embryo
Primary myogenesis: The first stage of muscle fiber formation
Proprioception: The transmission of sensory information from a muscle to the central nervous system
Proteasome: A protein complex involved in the degradation of damaged proteins
Ragged red fiber: A pathological muscle fiber characterized by numerous mitochondria along its outer edge
Rapsyn: A membrane protein that interacts directly with acetylcholine receptors
Receptor: A molecule, either membrane-bound or cytoplasmic, that interacts with a ligand
Rigor mortis: The muscular stiffness that occurs a short time after death
ROS (reactive oxygen species): Molecules that can potentially damage cells during the aging process
Ruffini ending: A deep pressure receptor
Ryanodine receptor: A molecule in the endoplasmic reticulum that is responsible for the release of Ca^{++}
Sarcolemma: The plasma membrane covering a muscle fiber
Sarcomere: The fundamental contractile unit within a muscle fiber
Sarcoplasmic reticulum: The highly modified endoplasmic reticulum responsible for Ca^{++} storage and release in a muscle fiber
Satellite cell: A mononuclear stem cell located between a muscle fiber and its basal lamina
Scatter factor: A growth factor involved in the migration of muscle precursor cells between a somite and limb bud
Schwann cell: A modified glial cell surrounding peripheral nerve fibers
Scleraxis: A transcription factor involved in vertebral and tendon formation
Sclerotome: The portion of a somite that gives rise to vertebrae
Secondary myogenesis: The second wave of muscle fiber formation in the embryo
Sharpey's fiber: A collagen fiber, partially embedded in bone, that connects a tendon to bone
Shh (sonic hedgehog): A signaling molecule involved in many embryonic inductive processes
Signal transduction pathway: A molecular pathway connecting a surface receptor to the genetic material in the nucleus
Sole plate nucleus: A myonucleus underlying a neuromuscular junction
Somite: An embryonic mesodermal structure that give rise to bone, muscle, and dermis
Spasticity: Sustained muscle contractions often seen after denervation in humans
Spindle capsule: The connective tissue surrounding a muscle spindle
Synaptic vesicle: An acetylcholine-containing vesicle in a nerve terminal

T-tubule: An inward extension of the sarcolemma designed to bring the action potential more quickly into the interior of a muscle fiber
Tenascin: An extracellular matrix protein
Tendon: A connective tissue link between a muscle and its attachment
Tenoblast: A form of fibroblast in a developing tendon
Tenocyte: A mature fibroblast within a tendon
Terminal sprouting: The sprouting of new nerve processes in aging muscle
Testosterone: A steroid hormone involved in muscle growth and hypertrophy
Tetanic contraction: A sustained contraction of muscle
TGF (transforming growth factor): A powerful growth factor involved in many developmental and regenerative processes
Thick filament: A bundle of myosin molecules
Thin filament: A bundle of actin molecules
Titin: A large protein involved in maintaining sarcomeric structure
TNF (tumor necrosis factor): A cytokine involved in inflammation
Training: A disciplined activity designed to improve muscular strength and/or endurance
Transcription factor: A molecule that binds to DNA and influences gene expression
Triad: The combination of a t-tubule and two elements of the sarcoplasmic reticulum
Troponin: A protein found on thin filaments
Twitch: A single muscle contraction in response to a brief stimulus
Type grouping: The juxtaposition of several muscle fibers of the same type as the result of terminal sprouting of nerve fibers
Type I fiber: A muscle fiber characterized by containing slow myosin
Type II fiber: A muscle fiber characterized by containing fast myosin
Ubiquitin: A protein involved in the intracellular degradation of damaged proteins
Ubiquitin ligase: An enzyme involved in the production of a chain of ubiquitin molecules
Ubiquitin/proteasome pathway (UPP): A major metabolic pathway involved in the degradation of damaged proteins
Unipennate: A muscle without a central shaft from which muscle fibers originate
Upper motor neuron lesion: A lesion in the central nervous system that affects nerve fibers feeding into spinal motor neurons
VEGF (vascular endothelial growth factor): A growth factor heavily involved in the remodeling and sprouting of blood vessels
V_{o2max} (maximal aerobic capacity): A measure of the maximum utilization of oxygen by the body
Volkmann's contracture: A pathological condition characterized by sudden vascular insufficiency, leading to massive muscle damage
Wnt: A growth factor functioning in many developmental contexts
Z-line: A thin band located at either end of a sarcomere, connecting the thin filaments of adjacent sarcomeres

Index

'*Note:* Page numbers followed by "f" indicate figures, "t" indicate tables and "b" indicate boxes.'

H

I

L

M

Printed in the United States
by Baker & Taylor Publisher Services